Hamlyn all-colour pap

John

Chemistry

illustrated by
Peter Edwards

Hamlyn · London
Sun Books · Melbourne

FOREWORD

This book sets out to give a picture of the whole of the science of chemistry. Beginning with atomic theory, it goes on to describe the elements and their inorganic chemistry. Organic chemistry is introduced as a natural extension of the chemistry of carbon. Then follow two sections on analysis, and the book ends with a brief account of the more important chemical industries.

In such a small space it has obviously not been possible to be completely comprehensive. Instead, the book concentrates on the most important aspects of each topic, relating them wherever possible to modern practical applications. This is neither a text book nor an historical account of the science. There are no recipes of how to do experiments. It shows rather the present state of our knowledge and how chemistry fits in with and contributes to other natural and physical sciences.

J.O.E.C.

Published by the Hamlyn Publishing Group Limited
London · New York · Sydney · Toronto
Hamlyn House, Feltham, Middlesex, England
In association with Sun Books Pty Limited, Melbourne

ISBN 0 600 00124 5
Phototypeset by V. Siviter Smith & Co. Ltd.
Colour separations by Schwitter Limited, Zurich
Printed in Holland by Smeets, Weert

CONTENTS

ATOMS

Most objects around us seem completely solid–they are made up of such things as metal, glass, wood, or plastic. Even materials such as leather, cloth and paper are technically called solids to distinguish them from liquids such as water and gases such as air.

There are really only two ways that matter–solids, liquids and gases–can be made up. Either matter is *continuous* and made up of one complete piece like a sort of jelly. Or it is made up of tiny component pieces, rather in the way that a large house is made up of hundreds of similar bricks.

In fact, the second alternative is correct: all matter is made up of tiny particles called *atoms*. Consider sugar dissolving in water. Each sugar crystal consists of atoms and so does water, although it appears to be continuous. But there must be spaces between the atoms of water because that is where the atoms of sugar go when it dissolves.

According to this idea, called the atomic theory of matter, everything is made up of tiny atoms packed together; nothing is really solid and continuous. It is not a particularly new theory. It was suggested well over 2,000 years ago by the ancient Greek philosopher Democritus. He invented the name *atom* which comes from a Greek word meaning *that which cannot be cut*. In other words, an atom is the smallest piece of matter that can exist on its own; it cannot be split by ordinary means into anything smaller.

How big is an atom?

According to the atomic theory, everything consists of atoms with empty spaces between them. Using ingenious laboratory experiments and calculations, scientists have been able to find the sizes of atoms. They turn out to be extremely small–far too small to be seen even with the most powerful microscopes. It would take nearly a million atoms laid side by side to equal the thickness of this paper page.

(Opposite page) Molecules of a liquid, such as water, cannot pass between the tightly packed atoms of a solid, such as copper.

Water
Copper
Copper and water

Inside the atom

In 1897, the British physicist J. J. Thomson was studying cathode rays–the rays that are used today in television tubes to produce the picture on the screen. His experiments showed that these rays are actually tiny particles of matter that come from inside the atom. He showed that they are negatively charged and that they weigh about one two-thousandth as much as do the lightest atoms.

If these particles, called *electrons*, make up only a two-thousandth of an atom, what makes up the rest and how are the various pieces arranged? The answers to these questions were provided by the New Zealand born physicist Ernest Rutherford. As a result of experiments in which he fired beams of alpha particles through thin sheets of gold foil, Rutherford proposed that every atom has a central solid core called a *nucleus* round which the much smaller electrons orbit. Later researchers discovered that even the

nucleus is not indivisible and that it, too, is composed of smaller fundamental particles called *protons* and *neutrons*. Every atom is made up of three kinds of sub-atomic particle: electrons, protons, and neutrons. The protons and neutrons crowd together in the nucleus, and the electrons move round in their orbits.

All substances in the world–indeed in the whole Universe –are made up of one or more *elements*. A wrought-iron gate is made from the single element iron. Other things around us are made up of more than one element. For example, the air we breathe consists of a *mixture* of gases, each of which is an element. Finally there are substances

(Opposite page) Using a magnetic field to bend beams of positively charged ions in a discharge tube, J. J. Thomson discovered the isotopes of neon.

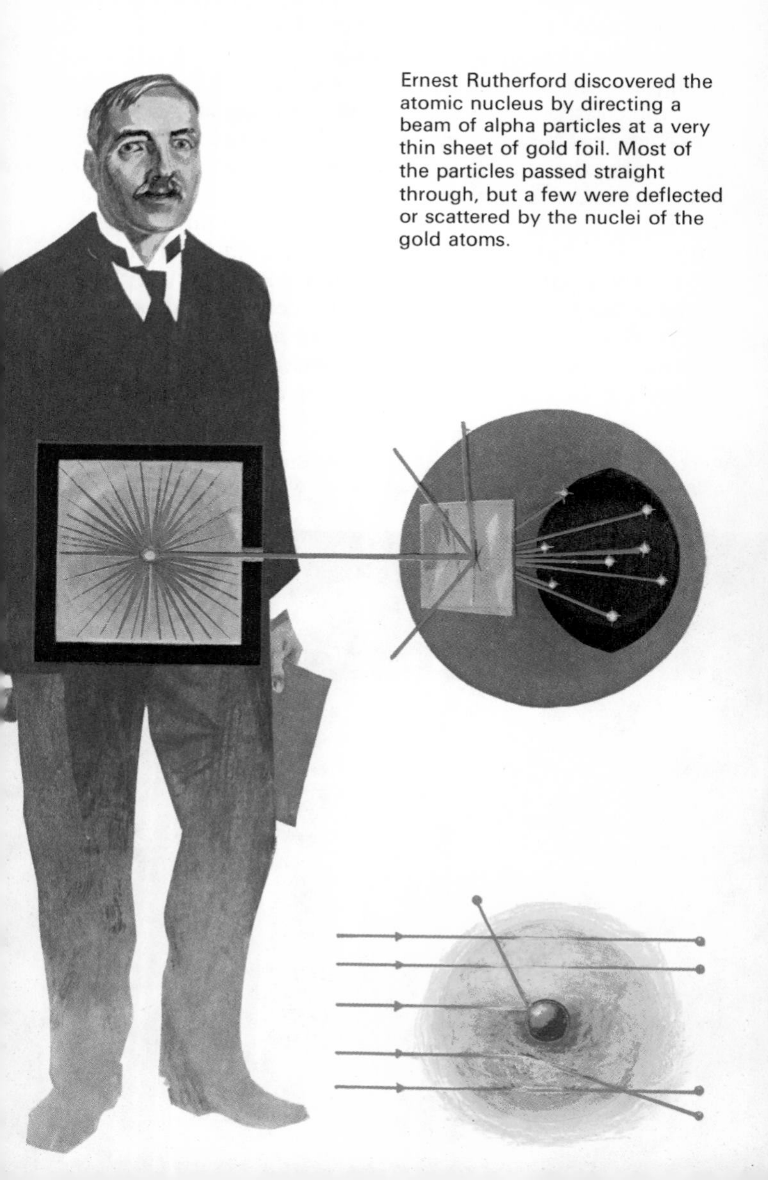

Ernest Rutherford discovered the atomic nucleus by directing a beam of alpha particles at a very thin sheet of gold foil. Most of the particles passed straight through, but a few were deflected or scattered by the nuclei of the gold atoms.

such as salt and sugar which are *compounds* consisting of more than one element in chemical combination with each other.

At the last count there were about 103 different elements known. But what makes an atom of, say, nitrogen different from an atom of oxygen? The answer to this question lies with the electrons, protons, and neutrons inside the atoms. Nitrogen has seven electrons and seven protons in its atoms, whereas oxygen has eight of each. The numbers of electrons and protons in the atoms of a single element are always the same and scientists give this figure a special name, the *atomic number*. So the atomic number of nitrogen is seven and that of oxygen is eight. But every different element has a different number of electrons and protons; all 103 elements have different atomic numbers.

What about the neutrons mentioned earlier? The first thing to remember is that they have no effect whatsoever on

According to the model proposed by Niels Bohr, an atom consists of a central nucleus surrounded by one or more orbiting electrons.

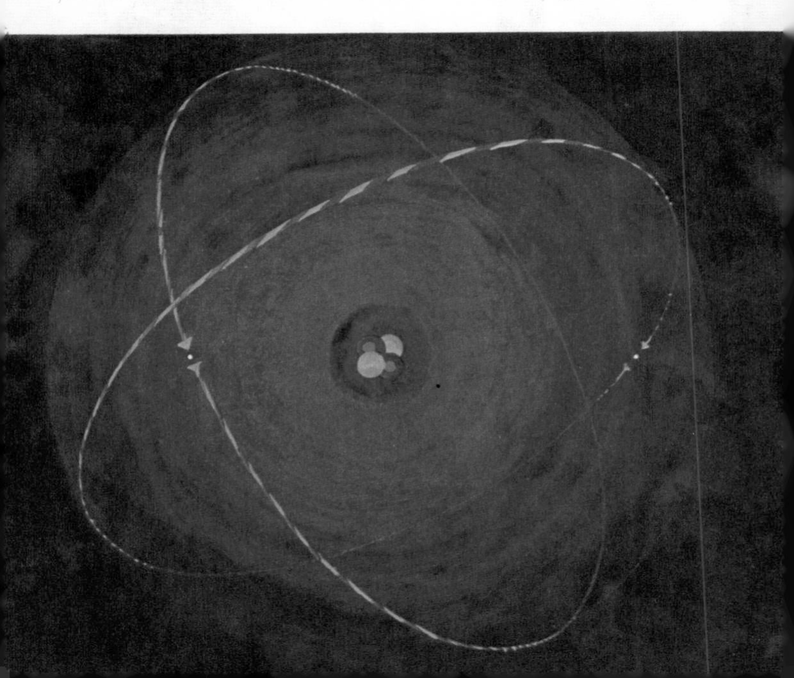

the ordinary chemistry of an element. As their name suggests, neutrons are electrically neutral and do not influence the chemical properties of an element one way or the other. But they do affect the mass of an element. If the masses of all the particles in an atom are added, a figure called the *atomic weight* or *atomic mass* is arrived at. The masses of a proton and a neutron are almost exactly one atomic mass unit.

Isotopes

About thirty years ago, a radioactive element of atomic mass fourteen was discovered. At first, it was believed to be nitrogen, which has the same atomic mass. But when its chemistry was studied, it was found to react exactly like the element carbon, which normally has an atomic mass of twelve. When the atomic *number* of the new radioactive element was determined, it was found to be six–exactly the same as carbon. Ordinary carbon has six protons and six neutrons in its nucleus, adding up to an atomic mass of twelve; a form of carbon had been discovered with six protons but an atomic mass of fourteen. This special carbon must have an extra two neutrons in its nucleus to bring the atomic mass up to fourteen. Such special forms of elements are called *isotopes:* they have the same number of protons as the ordinary element but a different number of neutrons and hence a different mass.

A mass spectrometer uses a powerful magnetic field to split a beam of charged ions and separate them according to their atomic masses.

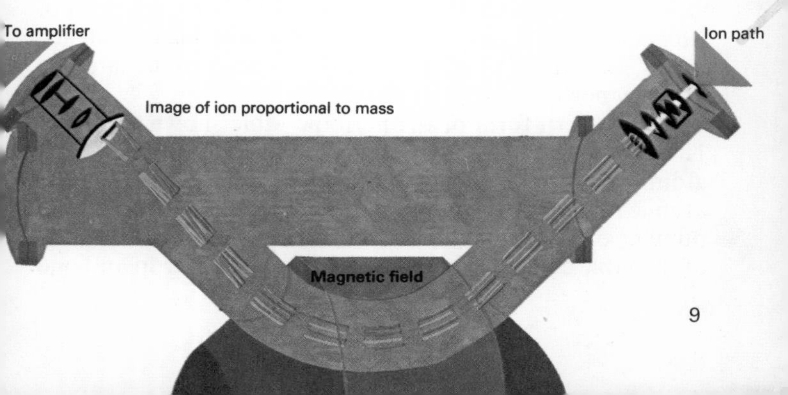

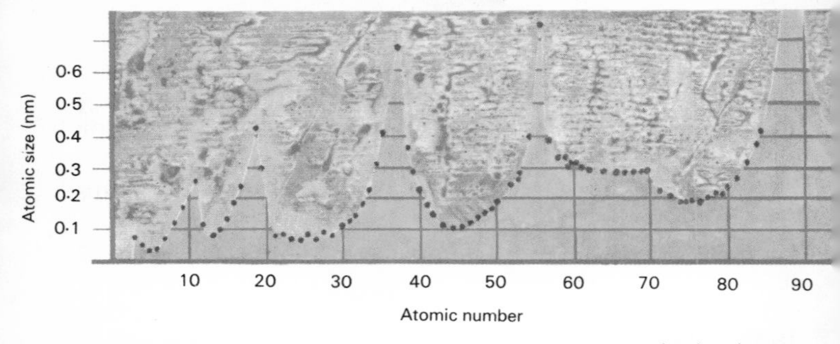

When a physical property of the elements, such as atomic size, is plotted against atomic number, the resulting graph has a series of peaks and valleys, showing a periodicity in the properties of the elements.

The periodic table

If the elements are listed in order of increasing atomic number, a definite pattern emerges: near the top of the list each element resembles the one appearing nine places farther down. This periodic recurrence in chemical properties was first spotted by the British chemist John Newlands in 1864, who arranged the elements in order of increasing atomic weight. A similar periodicity, this time in terms of physical properties, was noted by Julius Lothar Meyer in 1872.

But the great contribution to chemistry was made in 1869 by the Russian chemist Dmitri Mendeleeff who classified the elements into a table called the *periodic table*. He arranged elements of similar properties into columns (called groups), with horizontal rows (called periods) in order of increasing atomic weight. Whenever this treatment made an element fall into the 'wrong' group, Mendeleeff left a gap. He showed his genius and confidence in his system by predicting that the gaps would later be filled by newly discovered elements.

The modern form of the periodic table shown on pages 12 and 13 has the elements arranged in order of increasing atomic number (not atomic weight). Remember that the atomic number is the number of electrons in an atom or the number of protons in its nucleus. Each successive element of the table can be 'formed' by adding one electron and one proton, along with a few neutrons, to the element on its left. This is why one or two pairs of adjacent elements break the rule about increasing atomic weight. Iodine in Group VIIB follows tellurium, although iodine's atomic weight (126·9) is *less* than that of tellurium (127·6). This is because each element occurs naturally as a mixture of isotopes, and the heavy isotopes of tellurium are more abundant the light ones. Mendeleeff instinctively put the elements in the correct order to make iodine fall in the same group as its other halogen relatives. But the atomic number of iodine is fifty-three and that of tellurium is fifty-two, so all is well when atomic numbers are used instead of atomic weights.

Two other obvious features at the foot of the table are the two series of elements labelled lanthanides and actinides. The lanthanides are a set of fifteen elements of atomic numbers fifty-seven to seventy-one which slot into a single space in Group IIIA. They all have extremely similar chemical properties and are also known collectively as the rare earths. A similar series beginning with actinium (atomic number eighty-nine), called the actinides, contains uranium and all the man-made radioactive elements.

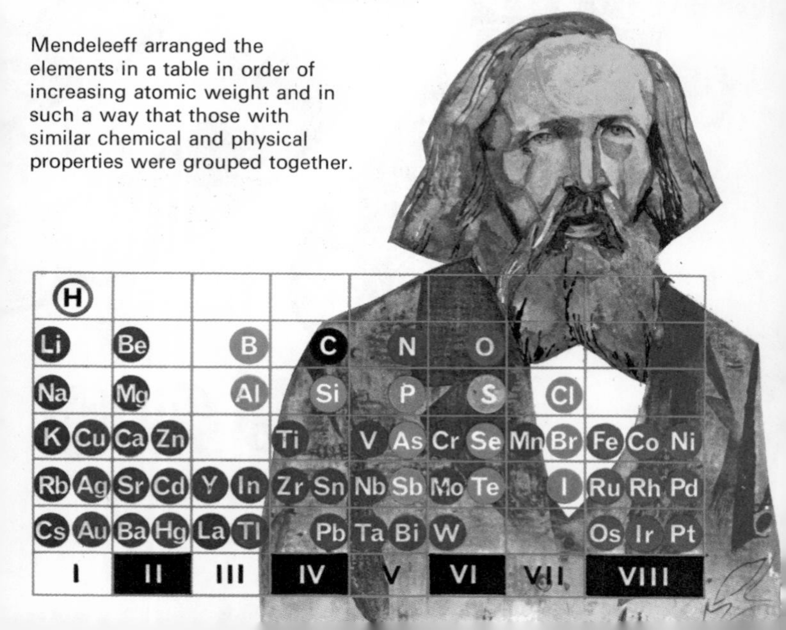

Mendeleeff arranged the elements in a table in order of increasing atomic weight and in such a way that those with similar chemical and physical properties were grouped together.

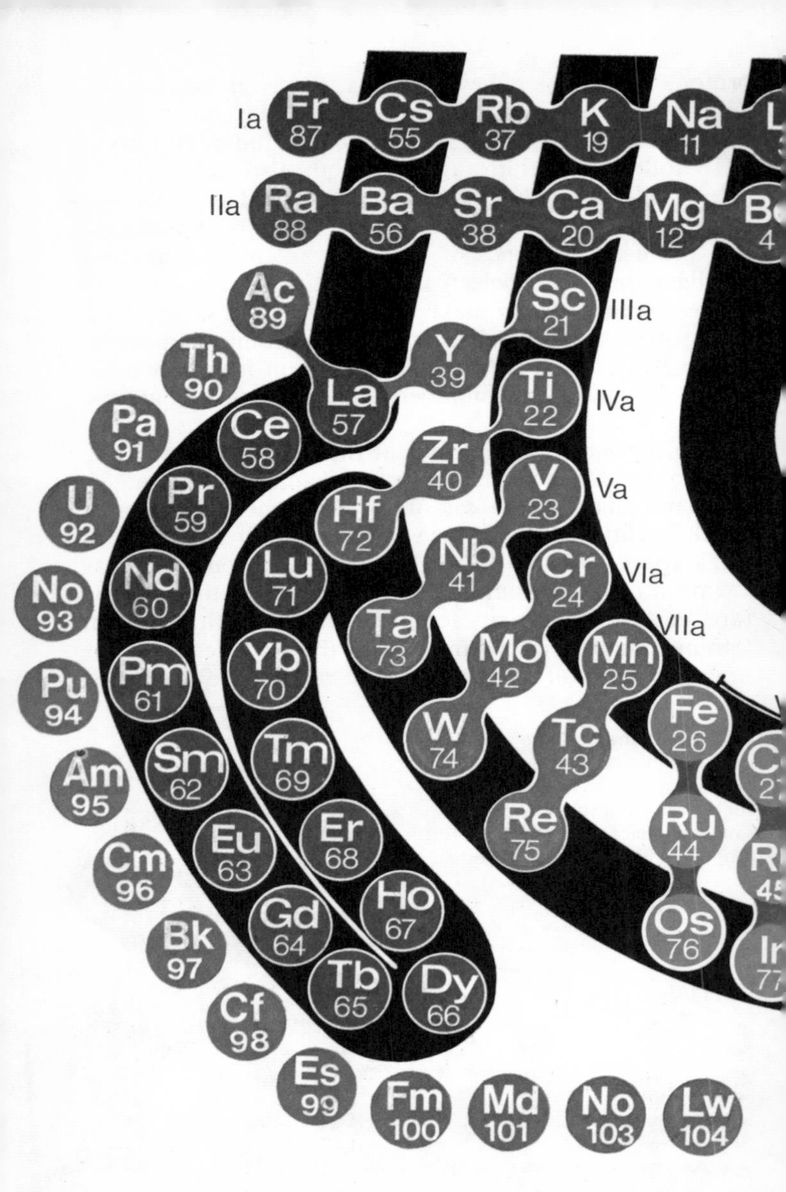
Ia
IIa
IIIa
IVa
Va
VIa
VIIa
Fr 87
Cs 55
Rb 37
K 19
Na 11
Ra 88
Ba 56
Sr 38
Ca 20
Mg 12
Ac 89
Sc 21
Y 39
La 57
Th 90
Ti 22
Pa 91
Ce 58
Zr 40
U 92
Pr 59
Hf 72
V 23
Nd 60
Lu 71
Nb 41
Cr 24
No 93
Ta 73
Pu 94
Pm 61
Yb 70
Mo 42
Mn 25
Am 95
Sm 62
Tm 69
W 74
Tc 43
Fe 26
Cm 96
Eu 63
Er 68
Re 75
Ru 44
Bk 97
Gd 64
Ho 67
Os 76
Tb 65
Dy 66
Cf 98
Es 99
Fm 100
Md 101
No 103
Lw 104

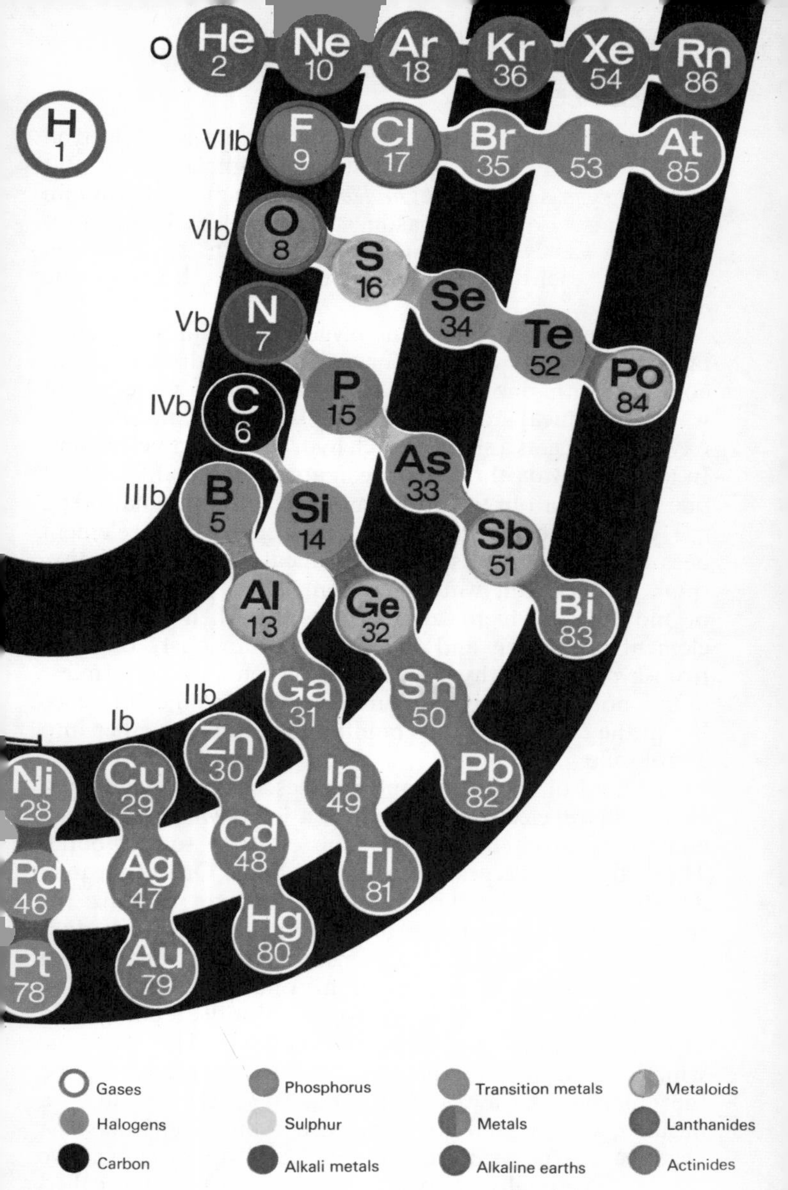
O
He 2
Ne 10
Ar 18
Kr 36
Xe 54
Rn 86
H 1
VIIb
F 9
Cl 17
Br 35
I 53
At 85
VIb
O 8
S 16
Se 34
Te 52
Po 84
Vb
N 7
P 15
As 33
Sb 51
Bi 83
IVb
C 6
Si 14
Ge 32
Sn 50
Pb 82
IIIb
B 5
Al 13
Ga 31
In 49
Tl 81
IIb
Zn 30
Cd 48
Hg 80
Ib
Cu 29
Ag 47
Au 79
Ni 28
Pd 46
Pt 78
Gases
Halogens
Carbon
Phosphorus
Sulphur
Alkali metals
Transition metals
Metals
Alkaline earths
Metaloids
Lanthanides
Actinides

MOLECULES

Hydrogen is the simplest atom. It has a single electron orbiting round a nucleus consisting of a single proton. But hydrogen gas is made up of *molecules*, each of which contain two atoms. For this reason, the chemical formula of hydrogen gas is H_2 and not simply H. In the hydrogen molecule, each nucleus effectively shares both of the electrons.

When hydrogen burns in oxygen, a chemical reaction takes place and water is formed. Two hydrogen atoms combine with one oxygen atom to form a molecule of water, chemical formula H_2O. In this molecule, the oxygen atom gets a share of each hydrogen atom's electrons. In a similar way, three hydrogen atoms will combine with one nitrogen atom to form a molecule of ammonia, NH_3.

Hydrogen, consisting of H_2 molecules, is an element because it cannot be split by chemical methods into anything simpler. But water is a compound, also consisting of molecules, which can be split into their constituent elements hydrogen and oxygen. It is obviously different from a mixture of hydrogen and oxygen (it is a harmless liquid, not an explosive mixture of gases), and the difference lies in the chemical bonds binding the atoms together into a molecule.

The kind of chemical bond just described, in which two atoms share electrons, is called a *covalent* bond. Each combining atom contributes one electron to the bond. There are other types of chemical bonds. One atom may donate both the binding electrons to form a *dative* (or *semipolar*) bond. But one of the commonest bonds does not involve a sharing of electrons. Instead, all the bonding electrons stay more or less round one of the combining atoms. This is the case in hydrogen chloride, a compound of hydrogen and chlorine of formula HCl. The electrons, which are negatively charged, tend to congregate round the chlorine. As a result, the chlorine atom acquires a negative charge and the hydrogen atom, which loses a negative charge, acquires an equal and opposite positive charge. Electrically charged atoms of this type are called *ions*. The resulting chemical bond, in which the atoms are held together by strong attractive electrostatic forces, is called an *ionic* (or *polar*) bond.

Valence

Notice in the preceding examples that hydrogen and chlorine are each able to form only one chemical bond, whereas each oxygen atom forms two (in water, H_2O) and each nitrogen atom forms three (in ammonia, NH_3). These numbers are characteristic of these elements and are called their *valence*. The alkali metals sodium and potassium also have a valence of one; they are said to be monovalent. Metals of Group IIA of the periodic table, which includes magnesium and calcium, are divalent. The commonest trivalent metal is aluminium (Group IIIB). Among the non-metals, carbon (Group IVB) has a valence of four and phosphorus (Group VB) can have a valence of five. An element may have more than one characteristic valence.

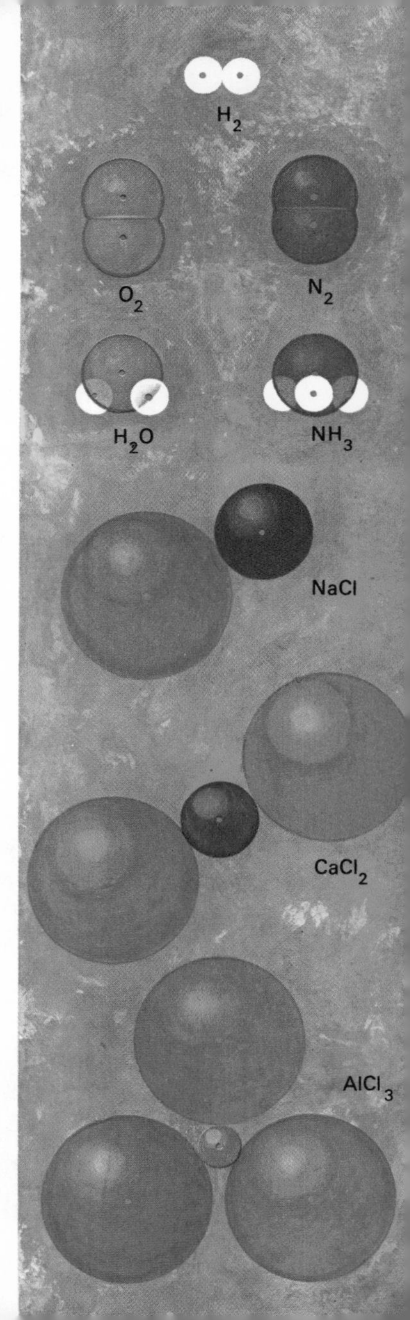

Various simple molecules, drawn approximately to scale.

Formulae and equations

There is no mystery about chemical formulae; they have been used already in this section. They are merely the chemist's shorthand way of writing down the composition of a molecule – that is, which atoms it contains and how many of each. The chemical formula H_2 for hydrogen gas indicates that each molecule of hydrogen contains two identical hydrogen atoms. Water, H_2O, consists of two hydrogen atoms bonded to one oxygen atom.

Some atoms are joined together in groups called *radicals*. These are not compounds because they cannot exist on their own, only in combination with other atoms or radicals. The carbonate radical is CO_3^{2-} and it is divalent. Chalk (calcium carbonate) is $CaCO_3$ and washing soda (sodium carbonate) is Na_2CO_3.

Chemists extend their shorthand to write down whole reactions. If sodium reacts with chlorine to form sodium chloride, the reaction can be written down in the form of an equation:

$$2Na + Cl_2 \rightarrow 2NaCl.$$

The symbols on the left-hand side of the equation stand for

A chemical formula is a shortened way of expressing the types and numbers of atoms in a molecule in terms of their chemical symbols.

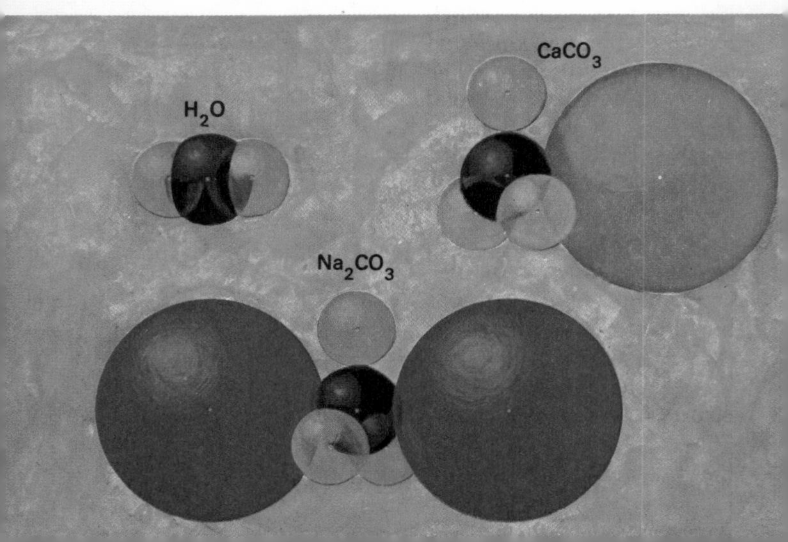

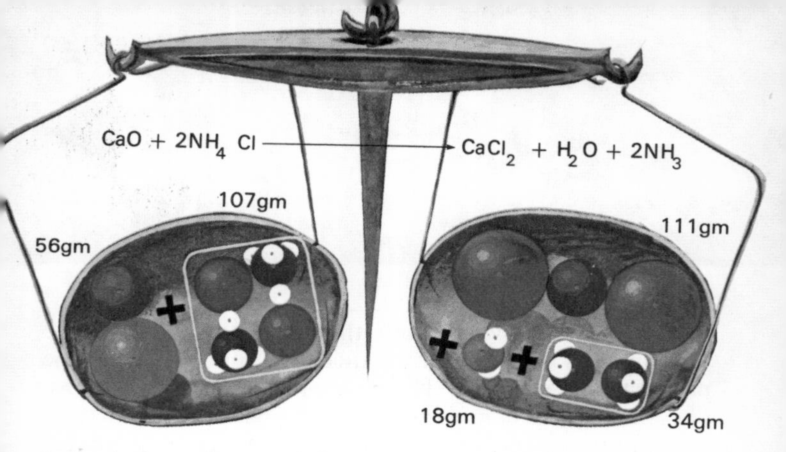

A chemical equation must balance—that is, the total weight of reactants (on the left-hand side) equals the total weight of products.

the reactants, those on the right-hand side for the products. Like any kind of mathematical equation, chemical equations have to balance–that is, there have to be the same numbers of sodium atoms and chlorine atoms on the left-hand side as there are on the right-hand side. The equation tells us that two atoms of sodium combine with a molecule of chlorine (consisting of two atoms) to form two molecules of sodium chloride (each consisting of two atoms).

Equations tell much more than what reacts with what–they can also indicate how much. To determine the quantities the atomic weights are written beneath the symbols of the elements concerned (using half quantities for simplicity):

$$\begin{array}{ccccc} Na & + & Cl & \rightarrow & NaCl. \\ 23 & & 35{\cdot}5 & & 58{\cdot}5 \end{array}$$

This shows that 23 parts by weight of sodium metal react with 35·5 parts by weight of chlorine gas to give 58·5 parts by weight of the salt sodium chloride. NaCl is the formula for a molecule of sodium chloride and 58·5 is its *molecular weight*. The molecular weight is the sum of the atomic weights of the component atoms. It may be expressed in any convenient units, such as grams or, if necessary, tons.

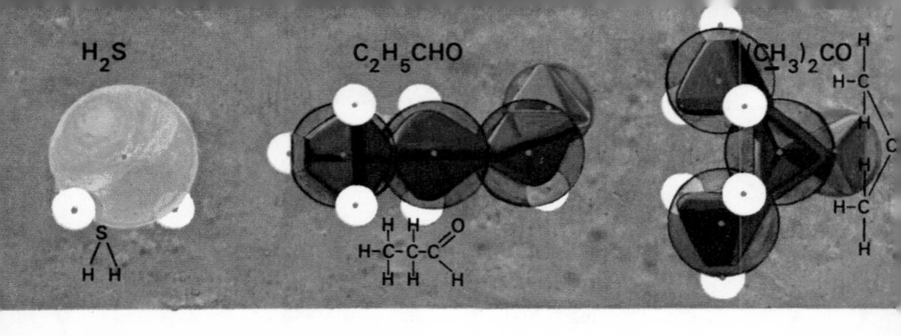

Conservation of mass

Notice in the above example that the total weight of the reactants (23 + 35·5 grams) equals the total weight of the products (58·5 grams). This reaction, as do all reactions, obeys the law of conservation of mass which states that, in a chemical reaction, matter is neither created nor destroyed.

Structural formulae

For certain purposes, chemists use a different kind of formula which gives, in addition to the numbers and kind of atoms in a compound, an indication of how the various atoms in a molecule are arranged. These structural formulae convey the same information as do the drawings of molecules on the previous pages, but they use the chemical symbols for the atoms concerned and do not require the chemist to have any artistic ability.

Structural formulae are particularly useful in organic chemistry, especially when two different compounds have the same atoms as each other, but differently arranged. Such substances are called *isomers*. The substances propionaldehyde and acetone both have the molecular formula C_3H_6O. They are, however, quite different, as their structural formulae show.

Stereochemistry

An extension of stating how atoms are arranged in molecules (that is, which atoms are joined to which) is to discover just where they are in relation to each other. This is the function of stereochemistry, which takes a three-dimensional look at molecules. It discovers how far apart atoms are and establishes the angles between chemical bonds. To the stereochemist, water is not merely a compound of

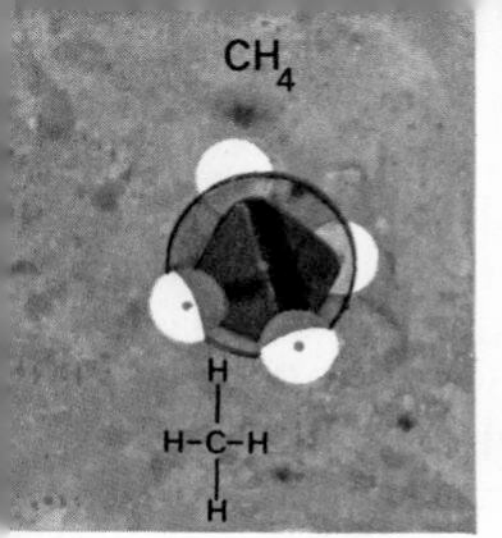

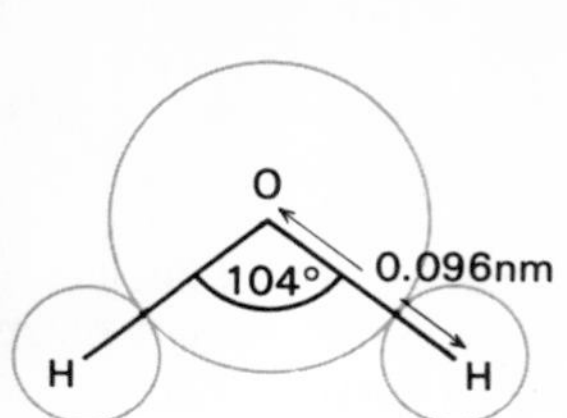

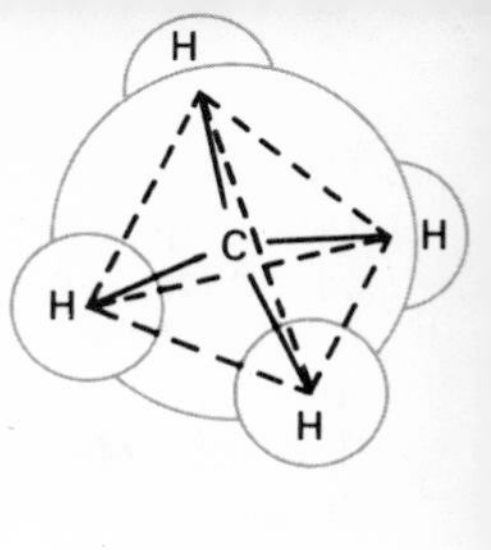

Molecular models may be made or drawn to show how the atoms are arranged in space. A water molecule is triangular, whereas a methane molecule is pyramidal with the four hydrogen atoms lying at the corners of a regular tetrahedron.

hydrogen and oxygen. It consists of molecules in which each hydrogen atom is 0·096 nanometers (1 nanometer is one ten millionth of a millimetre or 10^{-9} metres) from the oxygen atom, joined to it by bonds inclined at $104\frac{1}{2}°$.

Sometimes isomers exist when there is no immediately apparent reason why they should. For instance there are two substances called dichloroethylene both of which have the same chemical formula $ClCH_2{\cdot}CH_2Cl$. They have similar but not identical chemical and physical properties. Their structural formulae show that they both have the same atoms joined to each other in a similar order. But in one the two chlorine atoms lie on the same side of the carbon atoms, whereas in the other they lie on opposite sides. Two such substances are called *stereo* isomers, and in this example are *cis* (same side) and *trans* (opposite sides) forms of dichloroethylene.

Dichloroethylene molecules exist in two forms called stereo isomers. They differ in the positions of the chlorine atoms, which may be on the same side of the molecule (cis) or on opposite sides (trans).

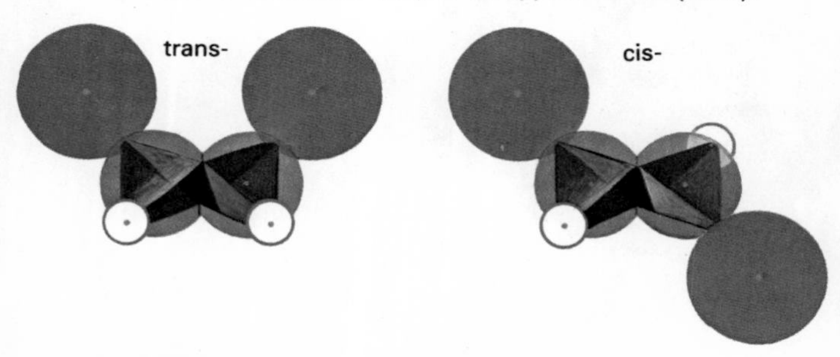

Gases, liquids, and solids

If the liquid water is cooled below 0°C, it freezes to a solid called ice. If water is heated above 100°C, it boils and turns into a gas called steam. Chemically ice, water, and steam are identical–they all consist of molecules of formula H_2O. The difference lies in the way the molecules are arranged. In ice, they are held rigidly in a crystal lattice, which is a sort of scaffolding of chemical bonds with molecules at all the junctions. Although each molecule can vibrate slightly, it never strays far from its proper place in the lattice. That is why a solid holds its shape.

When a solid is heated, the extra heat makes the molecules vibrate more vigorously. At first the effect is to make the molecules take up more room; that is why most solids expand on heating. Eventually the molecules vibrate so

Molecules in a solid, such as ice, have a close-packed regular arrangement. In a liquid such as water they move freely farther apart, and in a gas or vapour such as steam they move rapidly even farther apart.

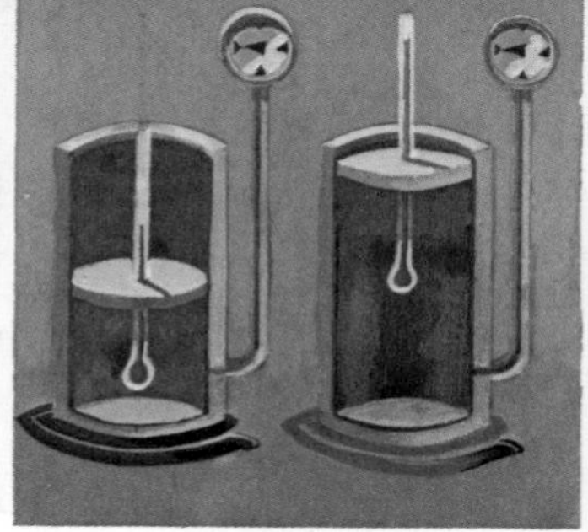

(Left) Boyle's law: at constant temperature, the volume of a gas is inversely proportional to pressure.
(Right) Charles's law: at constant pressure, the volume of a gas is proportional to temperature.

much that they abandon their positions in the lattice. The strength of the bonds between molecules is not sufficient to hold them in place, but there is still enough attraction between the molecules to prevent them flying apart altogether. When this happens, the solid melts and becomes a liquid.

If heating is continued, the molecules of the liquid vibrate more and more vigorously. At first, this has the effect of making the liquid more runny or, as the physicist would say, less viscous. Eventually the molecules are bounding around so much they leap clear of the surface of the liquid. The liquid boils and changes into a gas or vapour.

Gas molecules are constantly colliding with each other and with the walls of their container. The combined effect of millions of such collisions makes itself felt as the pressure exerted by the gas. If a container of gas is heated, the molecules vibrate more vigorously, collide more frequently, and the pressure of the gas rises. For the pressure to remain constant, the gas must expand. On the other hand, if more gas is forced into the container while keeping its temperature constant, there will still be more collisions (because there are more molecules) and the gas pressure again rises.

These effects caused by changing the temperature or volume of a gas were discovered experimentally long before scientists had any understanding of atoms and molecules. The discoverers were J. A. C. Charles and Robert Boyle who formulated laws which bear their names.

ACIDS AND ALKALIS

The gas hydrogen chloride, HCl, has a polar chemical bond joining its two atoms. The atoms exist as ions, the hydrogen ion carrying a positive charge and the chlorine ion a negative charge. When hydrogen chloride dissolves in water, it forms hydrochloric acid. The ions become almost entirely separated:

$$HCl \rightarrow H^+ + Cl^-.$$

This is true of all acids, which may be defined as substances that generate hydrogen ions in solution.

In strong acids, the hydrogen can be replaced by a metal. When this happens, the hydrogen is liberated as a gas and the metal forms a salt. For example, when zinc dissolves in hydrochloric acid the salt zinc chloride is formed:

$$Zn + 2HCl \rightarrow ZnCl_2 + H_2.$$

An alkali such as sodium hydroxide, NaOH, also consists of ions. In this case there are positively charged sodium ions and negatively charged hydroxyl ions which are also set free when the compound dissolves in water. An alkali may be defined as a generator of hydroxyl ions in solution:

$$NaOH \rightleftharpoons Na^+ + OH^-.$$

A fundamental reaction of an acid may be generalized by the statement: 'acid plus metal gives salt plus hydrogen'. Here iron reacts with sulphuric acid to form a solution of ferrous sulphate and to liberate hydrogen gas.

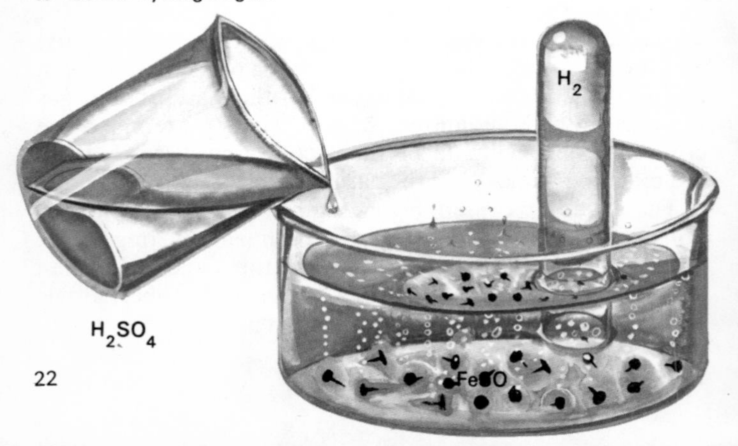

Strong alkalis are caustic; they cause painful burns on the skin because they can dissolve flesh. Hydroxyl ions from an alkali will combine with the hydrogen ions from an acid to form water:

$$H^+ + OH^- \rightarrow H_2O.$$

At the same time, the non-hydroxyl part of the alkali (generally a metal ion) and the non-hydrogen part of the acid (generally a non-metal ion) come together to form a salt. For example, the equation for the reaction between hydrochloric acid and sodium hydroxide is:

$$HCl + NaOH \rightarrow NaCl + H_2O.$$

The salt formed is sodium chloride or common table salt. This gives us another way of defining an alkali: it is a substance that reacts with an acid to form a salt and water.

Consider the reaction in which magnesium oxide dissolves in sulphuric acid:

$$MgO + H_2SO_4 \rightarrow MgSO_4 + H_2O.$$

Here again the substances formed are a salt and water (magnesium sulphate is epsom salt). But magnesium oxide is not an alkali; it is a white solid, insoluble in water, and not in the least caustic. Another more general term is used for such a substance; it is called a *base*.

Indicators

Pickled cabbage turns bright red when it is stewed in vinegar because it contains a vegetable dye which changes colour in the presence of acids. Chemists call these dyes,

Chemists test for acids and alkalis with indicators, often impregnated in paper strips, which change colour in acid or alkali.

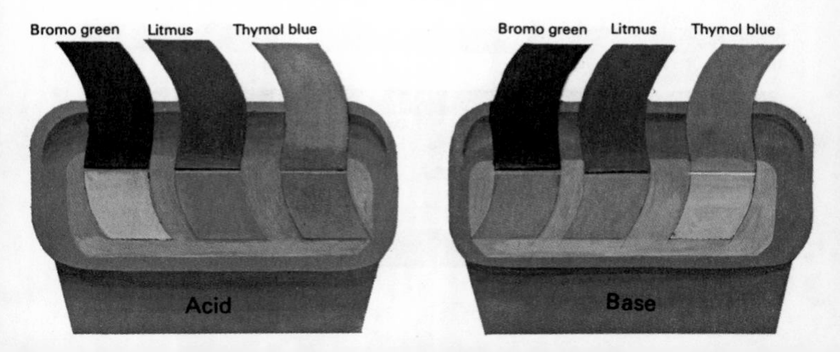

and their synthetic counterparts, indicators and use them to distinguish between acids and alkalis. A particularly useful natural indicator is litmus. It is red in acid solutions and blue in alkaline solutions.

Acidity and alkalinity

The strength of an acid depends on the number of hydrogen ions it releases in solution. Unfortunately these hydrogen ion concentrations are extremely small, varying from a tenth to a millionth of a gram ion per litre (a gram ion is the ionic weight, in the case of hydrogen 1, in grams). Small fractions such as these are often expressed in index form. For example, a tenth is 10^{-1}, a thousandth is 10^{-3}, and a millionth is 10^{-6}. Chemists use a scale of acidity called the *p*H scale which uses the index, neglecting the minus sign (mathematically, *p*H is the negative logarithm of the hydrogen ion concentration). For instance, a *p*H of 3 corresponds to an acid with 10^{-3} (a thousandth) gram ions of hydrogen per litre.

A neutral solution has a hydrogen ion concentration of 10^{-7} gram ions per litre, corresponding to a *p*H of 7. Alkaline solutions have a *p*H of more than 7. Because the *p*H scale is based on logarithms, a change of one unit on the scale corresponds to a change in acid or alkali concentration of ten times.

Salts

We have seen that whenever an acid is neutralized by an alkali a salt is formed. An acid also reacts with a metal to form a salt and hydrogen. These are just two ways of making

The strength of an acid or alkali, expressed as its *p*H, can be estimated by the colour of an indicator paper.

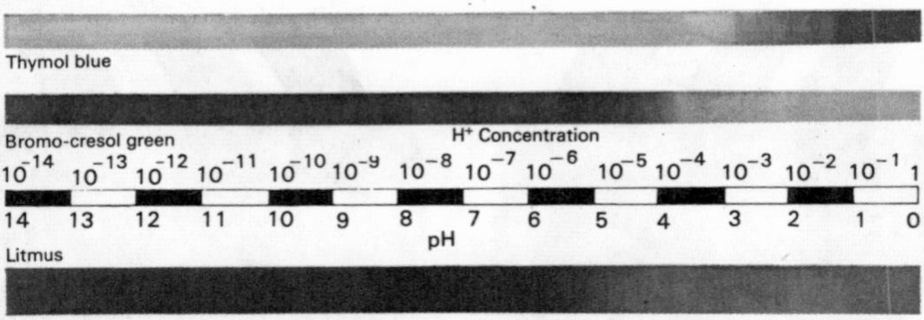

Thymol blue

Bromo-cresol green

H⁺ Concentration

10^{-14}	10^{-13}	10^{-12}	10^{-11}	10^{-10}	10^{-9}	10^{-8}	10^{-7}	10^{-6}	10^{-5}	10^{-4}	10^{-3}	10^{-2}	10^{-1}	1
14	13	12	11	10	9	8	7	6	5	4	3	2	1	0

pH

Litmus

salts. A few may be made by direct chemical combination between the elements. For example, magnesium chloride is formed when magnesium metal burns in chlorine gas:

$$Mg + Cl_2 \rightarrow MgCl_2.$$

When a soluble salt dissolves in water, it splits up into its ions:

$$NaCl \rightleftharpoons Na^+ + Cl^-.$$

This process is called *dissociation*. The double arrows tell us that this is a reversible reaction. When sodium chloride is crystallized from solution, the sodium ions and chloride ions come together again to form sodium chloride crystals. When two salts dissolve in water, they both dissociate. The various ions may then change partners, forming two new salts. If silver nitrate is added to a solution of sodium chloride, it first dissociates:

$$AgNO_3 \rightleftharpoons Ag^+ + NO_3^-.$$

(Left) A salt may be made by a direct combination of elements, as when magnesium burns in chlorine to form magnesium chloride. *(Right)* Salts are made by double decomposition when two other salts in solution 'change partners'; here potassium iodide reacts with lead nitrate to form potassium nitrate and a yellow precipitate of lead iodide.

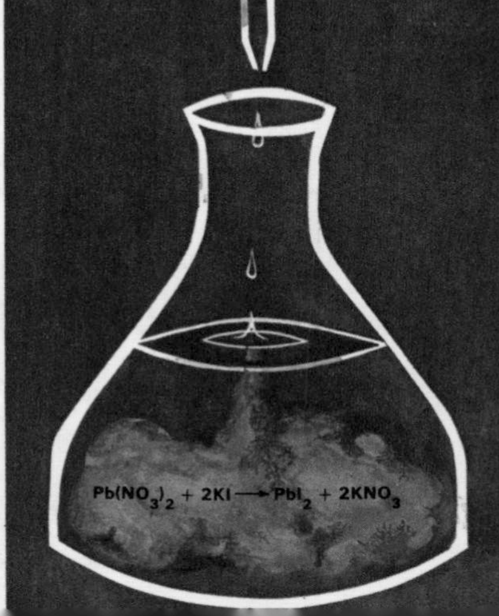

The solution contains ions of sodium, silver, chloride and nitrate. These change partners to form the two new salts sodium nitrate and silver chloride, because one of the salts–silver chloride–is insoluble. The overall equation for the reaction is:

$$NaCl + AgNO_3 \rightarrow NaNO_3 + AgCl.$$

This important way of making salts is called *double decomposition.*

Electrolysis

When an acid, an alkali, or a salt dissolves in water it dissociates into ions. If two metal plates, called electrodes, are dipped into such a solution and connected to a battery, the solution conducts electricity. A conducting solution of ions is called an electrolyte. The charged ions travel through the electrolyte carrying the electricity with them (*ion* comes from the Greek word for *traveller*) in exactly the same way as charged electrons carry electric current along a wire. The electrode connected to the negative terminal of the battery is called the *cathode* and the electrode connected to the positive terminal is the *anode*. Positively charged ions (called *cations*), such as H^+, Ag^+, and Cu^{2+}, move towards the cathode, whereas negatively charged ions (called *anions*), such as OH^-, Cl^-, and $SO_4{}^{2-}$, move towards the anode. When the ions arrive at the electrodes, they are dis-

Electroplating is a commercial application of electrolysis: the object to be plated is made the cathode (negative electrode) in a solution of a metal salt, such as copper sulphate.

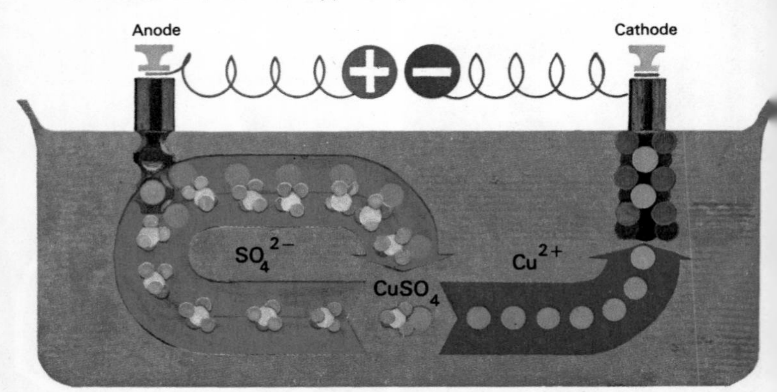

charged to form the corresponding atoms by addition or removal of electrons. In some cases, the liberated atoms appear as the free element–for instance, discharged hydrogen ions appear as hydrogen gas, discharged chloride ions appear as chlorine gas, and discharged silver ions appear as a deposit of metallic silver on the cathode. In other cases, the discharged ions promptly react with the water of the electrolyte to re-form ions.

Consider a solution of copper sulphate with copper electrodes connected to a battery. In the solution, the copper sulphate dissociates into ions:

$$CuSO_4 \rightleftharpoons Cu^{2+} + SO_4{}^{2-}.$$

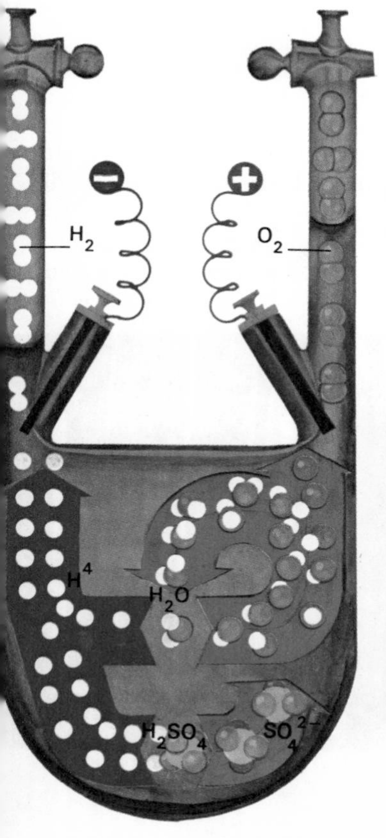

(Left) Electrolysis splits acidified water into its component elements, yielding two volumes of hydrogen at the cathode for each volume of oxygen at the anode.

(Below) In a solution, most metal salts are dissociated into their component charged ions; in electrolysis, the ions migrate to the electrode of opposite charge.

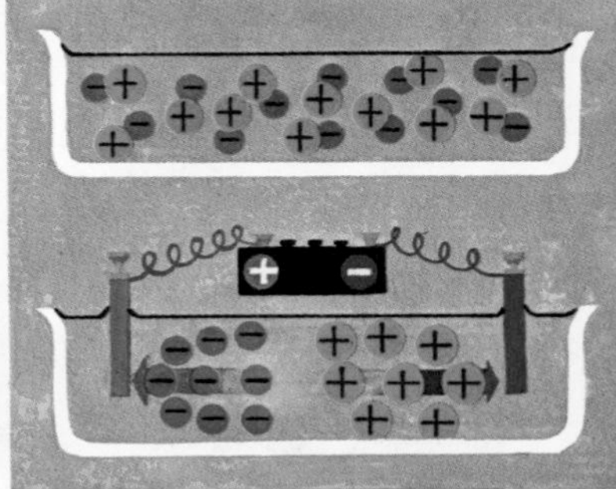

The Cu^{2+} cations move towards the cathode, and the $SO_4^{\,2-}$ anions move towards the anode. At the cathode, the Cu^{2+} ions are discharged by combining with electrons and form a thin film of copper metal on the electrode. This is the basis of a method of copper plating and of extracting pure copper from scrap metal (which is dissolved in sulphuric acid to make copper sulphate). The situation at the anode is a little more complicated. Instead of $SO_4^{\,2-}$ ions discharging, copper metal from the anode dissolves to form more Cu^{2+} ions leaving behind electrons. In this way, copper ions are formed at the anode as fast as they are discharged at the cathode and the concentration of these ions in the electrolyte remains constant. Also the electrons which are delivered by the battery at the cathode are effectively replaced by those which appear at the anode.

Batteries

In electrolysis, an electric current is used to make a chemical reaction take place. In a battery, or what is known as an electrolytic cell, a chemical reaction is used to generate electric current. An electric current consists of a stream of

Chemical reactions at electrodes generate electric currents in cells and batteries. The simple battery *(left)* soon becomes polarized, an effect which is overcome in the practical Daniell cell *(right)*.

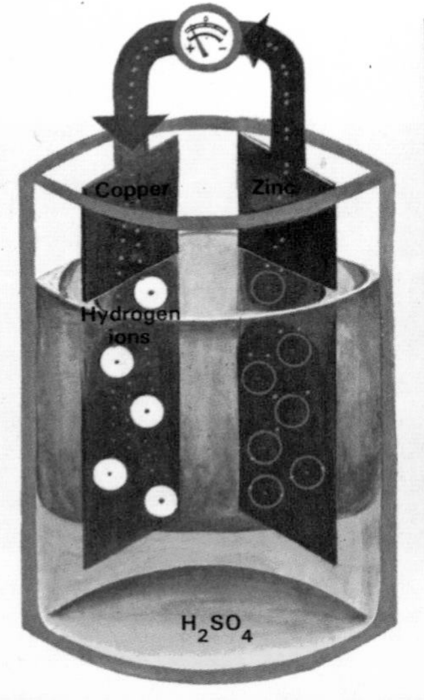

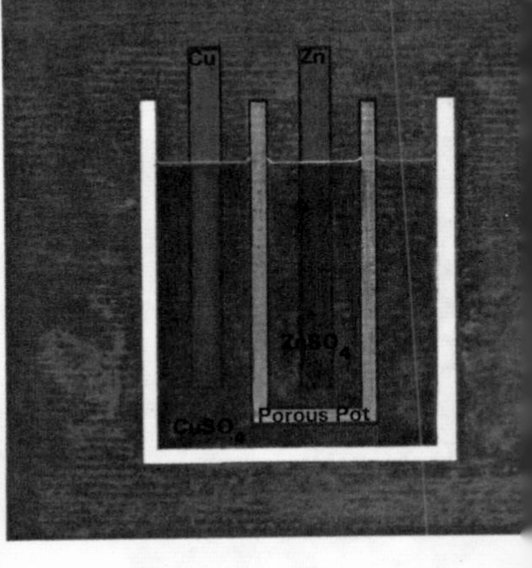

electrons flowing along a wire (or a stream of ions flowing across an electrolyte). A simple cell consists of a container of electrolyte with two different metal plates (also called electrodes) dipping into it. For example, electrodes of pure zinc and pure copper may dip into a solution of dilute sulphuric acid. When the plates are joined by a piece of wire, an electric current flows through the wire. What happens is that zinc tends to dissolve from the cathode and form zinc ions:

$$Zn \rightarrow Zn^{2+} + 2e.$$

The electrons formed are 'left' on the plate from where they flow along the wire as an electric current. The electrons then accumulate on the copper anode, where they pass back into the electrolyte by discharging hydrogen ions from the sulphuric acid. The discharged H^+ ions form hydrogen gas which is evolved at the anode.

After a while, such a simple cell stops generating current because the copper anode gets covered by a layer of hydrogen bubbles. This effect is called *polarization*. Working cells have a special arrangement to prevent polarization.

The Leclanché cell *(left)* has an ammonium chloride electrolyte and manganese dioxide as a depolarizer. The same chemicals are used in paste form in the ordinary dry battery *(right)*.

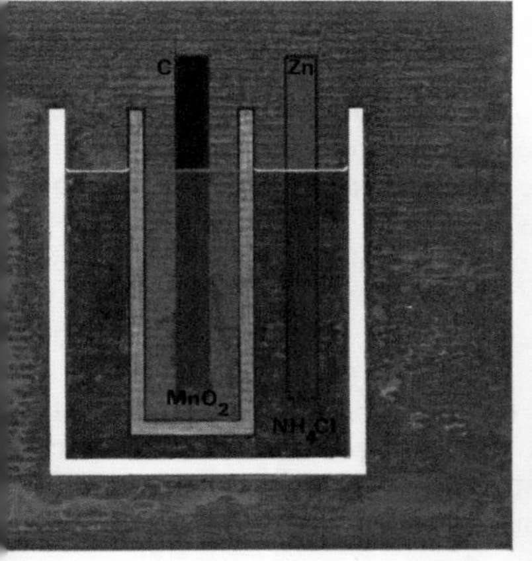

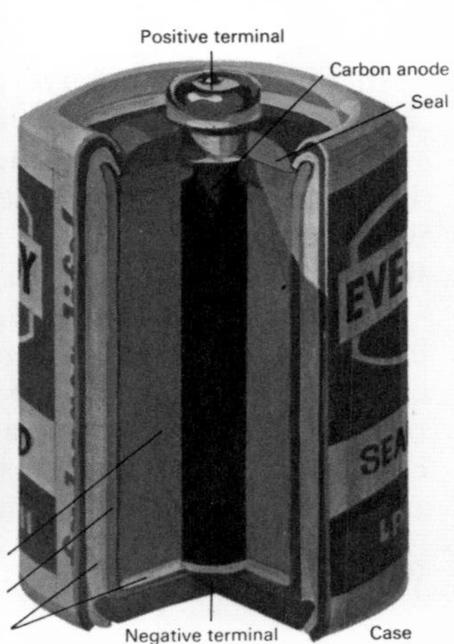

GASES

Eleven of the elements are gases, and these will be described in this section and the one following. Most of them belong to the first two periods of the periodic table.

Hydrogen

When a hydrogen bomb explodes, hydrogen atoms are brought together at extremely high temperatures. The hydrogen nuclei take part in a series of reactions in which they effectively fuse together in groups of four to form atoms of helium.

The hydrogen bomb, scaled up several billion times, is essentially the reaction that takes place in the Sun and other stars. Every second in the Sun, more than 550 billion tons of hydrogen fuse to form helium. Hydrogen is therefore the most important element in the universe, making up 98 per cent of it. On Earth it occurs almost entirely in the form of its compounds. Two-thirds of the Earth's surface is covered with water – the best-known compound of hydrogen. Carbohydrates such as sugar, starch, and cellulose contain hydrogen, as do hydrocarbons in fuels such as coal and oil.

To make hydrogen, it must be removed from one of its compounds. Commercially it is made from water by the Bosch process. Steam passes over red hot iron filings; the iron combines with the oxygen in steam, liberating hydrogen:

$$3Fe + 4H_2O \rightarrow Fe_3O_4 + 4H_2.$$

In the laboratory, hydrogen is generally made by the action of an acid on a metal. Hydrogen may be liberated violently from water by the action of an alkali metal such as sodium or gently by electrolysis. Calcium decomposes water less vigorously, and magnesium generates hydrogen from steam.

In addition to being used as a fuel, hydrogen is used commercially in the Haber process in which it is made to combine with nitrogen to form ammonia, the source of many fertilizers and explosives. In the presence of a suitable catalyst (a *catalyst* is a substance that speeds up a chemical reaction without itself taking part in it), hydrogen will com-

bine with oils to produce solid fats, such as margarine.

Hydrogen burns with a pale blue flame to form water. A mixture of hydrogen and oxygen combines explosively if sparked, and under control in a welding torch, hydrogen burns in oxygen to produce the extremely hot oxy-hydrogen flame.

Hydrogen, generally made in the laboratory by the action of dilute hydrochloric acid on zinc, was used in airships and is the 'fuel' in a hydrogen bomb.

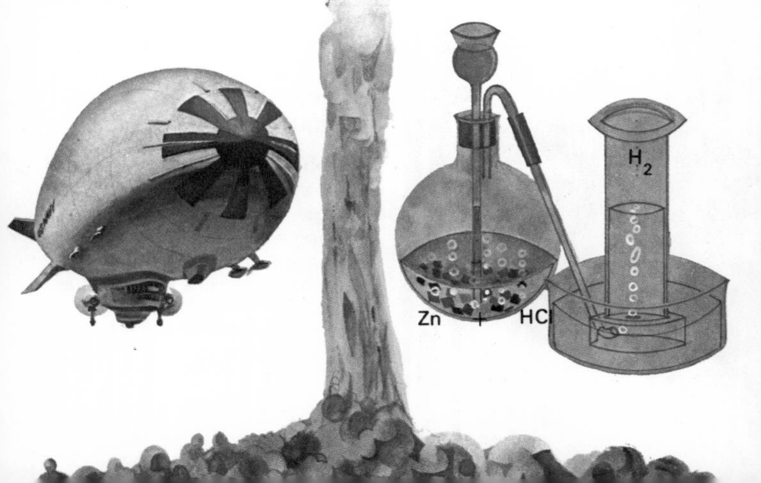

Oxygen

The most common element on Earth is oxygen. It makes up nearly half of the Earth's crust, combined as oxides and salts in rocks and as eight-ninths of all water by weight. Free oxygen gas forms about a fifth of the air, most of the remainder being nitrogen. It is breathed in by men and animals; it is also necessary for combustion, and practically all forms of burning involve a chemical reaction with oxygen.

Commercially oxygen is obtained from air. At low temperatures and under high pressures, air condenses to a liquid. When liquid air is allowed to warm up, the more volatile liquid nitrogen boils off leaving liquid oxygen behind.

In the laboratory, oxygen is made from one of its compounds. When potassium nitrate is strongly heated, it decomposes into the nitrite and oxygen:

$$2KNO_3 \rightarrow 2KNO_2 + O_2.$$

Oxygen may be made by heating a dry mixture of potassium chlorate and manganese dioxide. It is used as an oxidizer in liquid fuel rockets and is the vital component in the air we breathe.

A similar reaction takes place when potassium chlorate is heated. In this case, the process is generally speeded up by using manganese dioxide as a catalyst.

A method of historical importance in chemistry involves heating mercury oxide. This method was used by the Swedish chemist Karl Scheele when he discovered oxygen in 1769 and, quite independently, by Joseph Priestley in 1774.

Many non-metals and metals combine with oxygen to form oxides. Sulphur burns in oxygen to give the choking gas sulphur dioxide, SO_2, carbon forms carbon dioxide, CO_2, and phosphorus gives phosphorus pentoxide, P_2O_5. Each of these gases dissolves in water to form an acid, and for this reason they are called *acidic* oxides.

Alkali metals such as sodium and potassium burn in oxygen to give oxides that dissolve in water to produce alkalis. Such oxides are called *basic* oxides. Iron and copper burn in oxygen to produce oxides that do not dissolve in water. But since these oxides do dissolve in acids to form salts, they are also classified as basic.

Finally there are oxides such as water, carbon monoxide, and nitric oxide. These substances have neither predominantly acidic nor predominantly basic reactions and are called *neutral* oxides.

Oxygen is used for breathing by pilots, astronauts, and divers. Together with fuel gases such as hydrogen or acetylene, it is used for producing very hot flames for cutting metals and welding. More recently, oxygen has been used as a component for liquid-propellant rockets with fuels such as petrol or kerosene.

Ozone

Molecules of oxygen have the formula O_2 because they each contain two atoms. If a stream of air or oxygen is passed through an electric spark, atoms of oxygen combine in triplets to form the gas ozone, O_3. Ozone has a pungent acrid smell and occurs naturally in the upper atmosphere where it acts as a filter, stopping most of the Sun's harmful ultra-violet radiation from reaching the Earth. It is sometimes used as a disinfectant to kill germs in the air pumped into such places as underground railways.

Water

Water is formed whenever hydrogen or fuels containing hydrogen burn. Its composition can easily be demonstrated by electrolysis when hydrogen is evolved at the cathode and oxygen at the anode. If the gases are collected, the volume of hydrogen will be twice that of oxygen, confirming the well-known formula H_2O.

Water is also formed when a metal oxide is reduced to the metal by hydrogen:

$$CuO + H_2 \rightarrow Cu + H_2O,$$

or when an acid is neutralized by a basic oxide or by an alkali to form a salt:

$$ZnO + H_2SO_4 \rightarrow ZnSO_4 + H_2O,$$

$$NaOH + HCl \rightarrow NaCl + H_2O.$$

Hydrogen peroxide

Certain metal oxides react with acids to give hydrogen peroxide instead of water:

$$BaO_2 + 2HCl \rightarrow BaCl_2 + H_2O_2.$$

Hydrogen peroxide is used as a convenient source of oxygen for combustion, for bleaching, and for disinfecting.

Oxidation and reduction

All reactions in which oxygen combines with other elements are examples of oxidation. Substances such as potassium chlorate, ozone and hydrogen peroxide which readily make oxygen available are called *oxidizing agents*. The opposite process, in which oxygen is removed from a compound, is called reduction and substances that facilitate it are *reducing agents*. The extraction of copper metal from copper oxide is a reduction with hydrogen as the reducing agent:

$$CuO + H_2 \rightarrow Cu + H_2O.$$

The removal of hydrogen is also an oxidation, even if oxygen gas is not involved. For example, the production of hydrogen by the action of hydrochloric acid on zinc involves an oxidation of the zinc:

$$2HCl + Zn \rightarrow ZnCl_2 + H_2.$$

In each case, the oxidation of one reactant is accompanied by a reduction of the other.

Burning

An oxidation that takes place rapidly with the production of heat and possibly flame is called combustion or burning. When hydrogen or coal gas burns, there is a flame and heat. When iron burns in oxygen or coke burns in a carefully

The burning of a house or of pure hydrogen to form water are both examples of oxidation; the reverse process, such as the reaction between copper oxide and hydrogen to liberate pure copper, is called reduction.

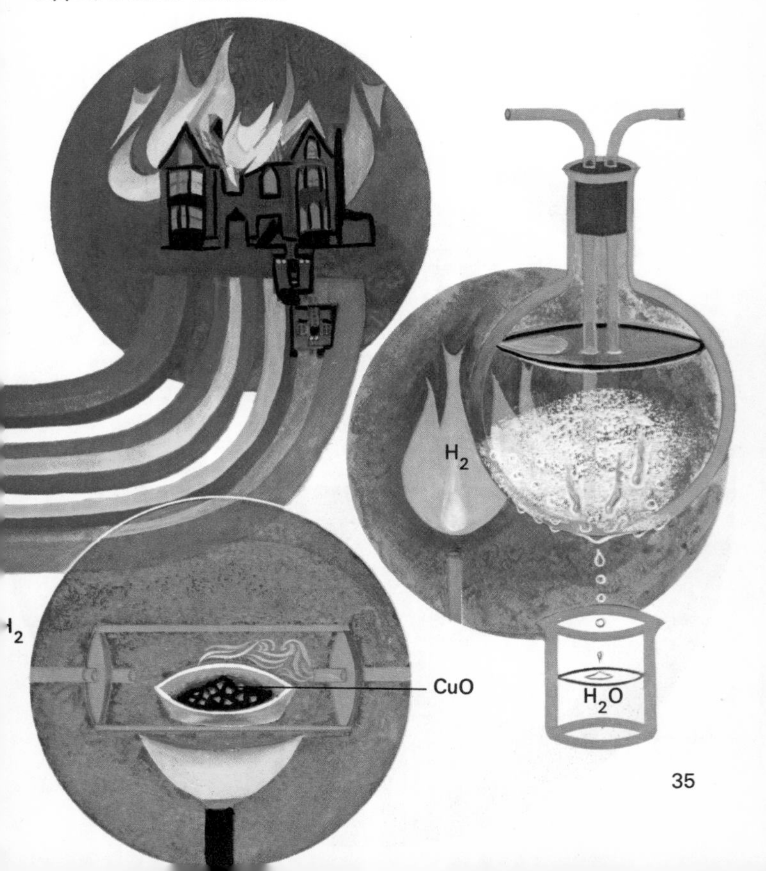

controlled draught, there is heat but no flame. In each case, oxygen combines with the burning substance (which is oxidized) to form oxides such as water, iron oxide, or carbon dioxide. To make a substance burn more vigorously, it must be supplied with extra air or oxygen. For this reason opening the air-intake below a dull coal fire makes it burn more brightly.

Nitrogen

Another element necessary to life is nitrogen, which is an essential part of proteins, the body builders of animal tissues. Plants also need nitrogen which is added to the soil in the form of fertilizers. Some plants of the pea family have on their roots nodules that house bacteria which can 'fix' nitrogen from the air and convert it into plant food-

The nitrogen cycle shows how nitrogen is stored and released by plants, animals and the atmosphere.

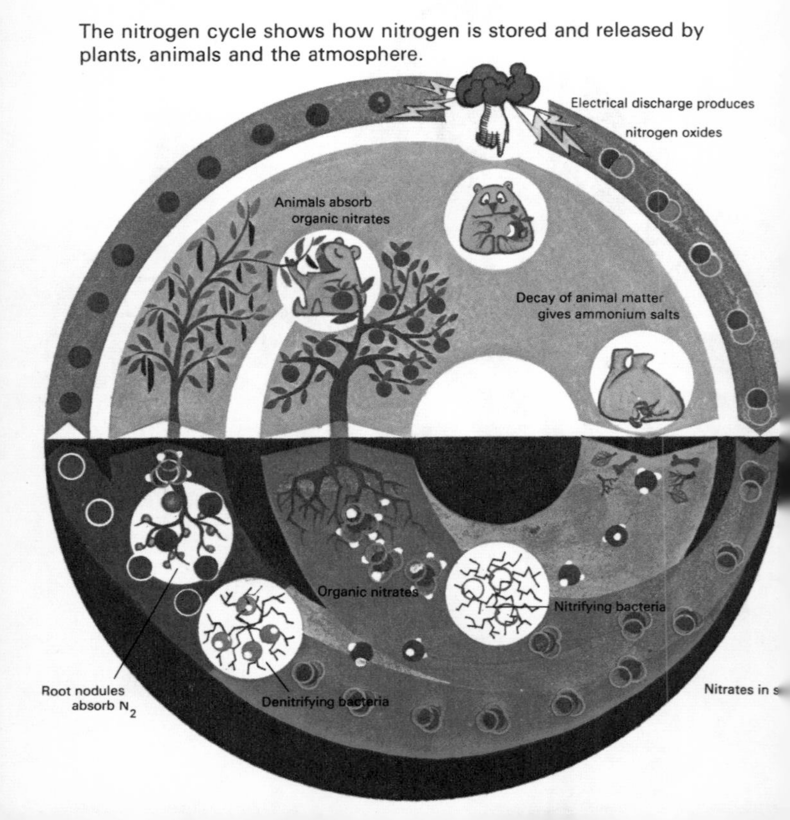

stuffs. A little nitrogen in the air is also fixed during thunderstorms by lightning, when it combines with oxygen to form nitric oxide. Nitrogen occurs as mineral deposits of nitrates in various parts of the world; potassium nitrate from these sources is called saltpetre.

Nitrogen is made commercially from air. Either air is liquefied and the nitrogen fractionally distilled off, or air is passed over hot coke (carbon) which combines with the oxygen to form carbon dioxide, leaving the nitrogen unchanged.

Nitrogen can be prepared in a similar way in the laboratory by passing air through a tube containing red hot iron filings to remove the oxygen. It is also produced when certain of its compounds are heated. Ammonium nitrate, for example, decomposes to water and nitrogen:

$$NH_4NO_2 \rightarrow N_2 + 2H_2O.$$

Two important industrial processes use nitrogen: in the Haber process, it combines with hydrogen to form ammonia, and in the Birkland-Eyde process nitrogen is passed through a huge electric spark (man-made lightning) to form nitric oxide for making nitric acid.

Ammonia

The Haber process for making ammonia uses the cheapest possible starting materials–water and air. The Bosch process (see page 30) produces hydrogen from the water, and nitrogen is extracted from the air by one of the methods just described. The reaction between hydrogen and nitrogen to form ammonia is incomplete and reversible:

$$3H_2 + N_2 \rightleftharpoons 2NH_3.$$

The problem of converting this difficult laboratory reaction into a workable industrial process was solved by the German chemist Fritz Haber immediately before the outbreak of World War I. In the Haber process, the gases are heated to about 500°C under 250 atmospheres pressure in the presence of an iron oxide catalyst. Ammonia is also produced as a by-product at the gas works.

In the laboratory, ammonia may be made by the action

of any hot alkali on any ammonium salt. For convenience chemists generally use lime as the alkali because it is a solid:

$$2NH_4Cl + CaO \rightarrow CaCl_2 + H_2O + 2NH_3.$$

Ammonium carbonate decomposes to give ammonia at room temperature and for this reason this compound is used in smelling salts.

Ammonia is an extremely soluble gas, dissolving to the extent of about 800 volumes in one volume of water. The resulting solution, called ammonium hydroxide, NH_4OH, behaves as an alkali and forms ammonium salts with acids. All ammonium salts are soluble in water and are used as fertilizers. On cooling to —33°C or under pressure at ordinary temperatures, ammonia condenses to a liquid which is used in refrigerators. But most commercially produced ammonia is oxidized by air with a platinum catalyst to form nitrogen oxides for making nitric acid.

Nitric acid

When nitric oxide, formed from the Birkland-Eyde process or from the catalytic oxidation of ammonia, is mixed with

(Below) Nitrogen, in the form of nitrates or ammonium salts, is used in artificial fertilizers. *(Opposite left)* The action of nitric acid on copper turnings generates colourless nitric oxide gas, which reacts with the oxygen in air to form brown fumes of nitrogen dioxide. *(Opposite right)* Nitrous oxide, or laughing gas, was one of the first anaesthetics.

air and dissolved in water, it forms nitric acid:

$$4NO + 3O_2 + 2H_2O \rightarrow 4HNO_3.$$

The acid is made in the laboratory, as it once was industrially, by the action of hot concentrated sulphuric acid on sodium nitrate:

$$NaNO_3 + H_2SO_4 \rightarrow NaHSO_4 + HNO_3.$$

Nitric acid is a strong oxidizing agent. With metals, instead of generating hydrogen it produces nitrogen oxides and water. It oxidizes non metals such as carbon and sulphur to their oxides and causes substances such as sawdust to catch fire. With many organic chemicals it forms explosives.

Nitrogen oxides

There are eight different oxides of nitrogen, the commonest being nitrous oxide, N_2O, nitric oxide, NO, and nitrogen dioxide, NO_2. Nitrous oxide–once called laughing gas–is sometimes used as an anaesthetic by dentists. It is made by cautiously heating ammonium nitrate:

$$NH_4NO_3 \rightarrow 2H_2O + N_2O.$$

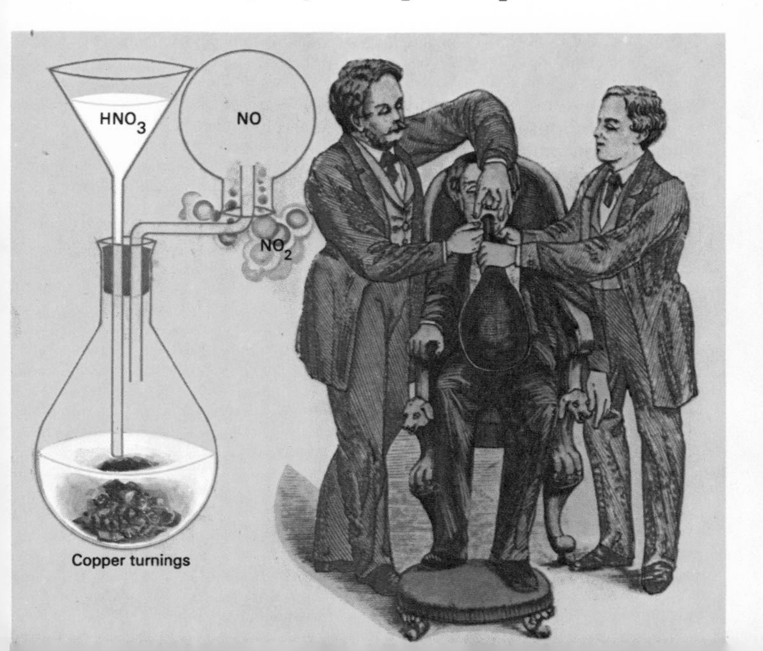

Nitric oxide is made by various industrial processes which are a preliminary to the manufacture of nitric acid. In the laboratory, it can be produced by adding 30 per cent nitric acid to copper turnings:

$$3Cu + 8HNO_3 \rightarrow 3Cu(NO_3)_2 + 4H_2O + 2NO.$$

Nitrogen dioxide is generally made in the laboratory by using copper turnings and *concentrated* nitric acid:

$$Cu + 4HNO_3 \rightarrow Cu(NO_3)_2 + 2H_2O + 2NO_2.$$

On cooling, molecules of nitrogen dioxide combine in pairs to form the colourless compound nitrogen tetroxide, N_2O_4.

The inert gases

During an eclipse of the Sun in 1868, the English astronomer Norman Lockyer pointed a spectroscope at the solar corona and spotted a new line in the spectrum. He ascribed the line to a new element which he christened helium (from the Greek *helios* meaning *Sun*). Twenty-seven years later, William Ramsay detected the same gas in uranium minerals. It was one of six such gases, and Ramsay had a hand in the

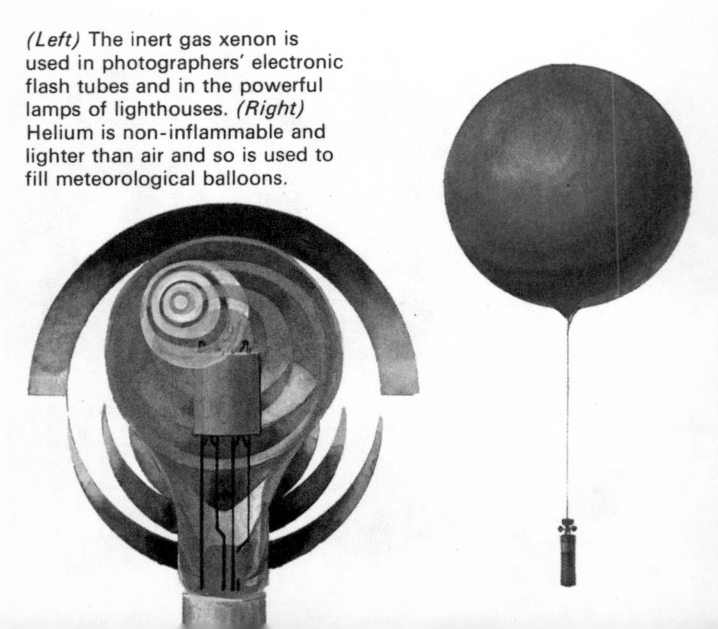

(Left) The inert gas xenon is used in photographers' electronic flash tubes and in the powerful lamps of lighthouses. *(Right)* Helium is non-inflammable and lighter than air and so is used to fill meteorological balloons.

discovery of all of them. They form a complete group (numbered 0) of the periodic table and, because they take part in practically no chemical reactions, they are called the inert gases.

When nitrogen is made from air by one of the methods described earlier in this section, it is found to be denser than nitrogen made from its chemical compounds. Lord Rayleigh noted this fact in 1894 and suspected it might be due to a previously undiscovered element in the atmosphere. Ramsay took up the hunt and discovered the first inert gas argon (named from the Greek word for *inert*). Argon makes up nearly one per cent of the air. It is used together with nitrogen for filling incandescent electric lamps.

Then in 1895 Ramsay discovered helium in uranium minerals. He also detected it in air and in natural gas from oil wells and mineral water springs. The alpha particles used by Lord Rutherford for splitting atoms are really helium nuclei, which accounts for the presence of helium in radioactive minerals. Helium also turns up in nuclear fusion reactions, hence its abundance in the Sun. Helium is the lightest gas next to hydrogen, and since it is non-inflammable it is used instead of hydrogen for filling balloons. Liquid helium boils at —268·9°C and is used for producing the lowest temperatures ever reached.

In 1898, Ramsay discovered three more inert gases in air: neon, krypton and xenon. Orange-red electric advertising signs contain traces of neon gas. Under similar conditions, krypton produces a greenish light. Xenon gives a bluish-white light and is used in photographers' electronic flash bulbs. It is the least inert of the group of gases, and compounds with fluorine and oxygen have been prepared.

Radon, the sixth inert gas, was discovered in 1901 by the German physicist Friedrick Dorn as a highly radioactive gas given off by radium, and identified as the last inert gas by Ramsay. Similar gases – now known to be lighter isotopes of radon – are given off by the radioactive elements actinium and thorium. All the isotopes eventually decay to lead, so that there is a situation where a metal spontaneously changes into a gas, which then changes back into a metal.

Many compounds of fluorine have commercial applications. Hydrofluoric acid is used for etching glass *(left)*. Organic compounds include Freon, used in refrigerators *(above)*, and the plastic PTFE, which forms the non-stick lining of modern frying pans.

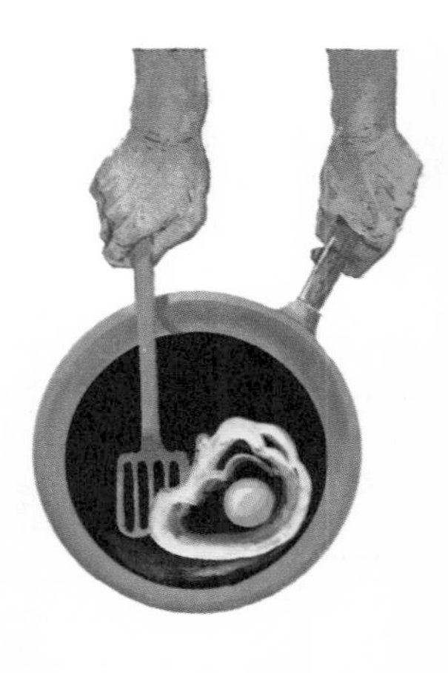

THE HALOGENS

The five elements of Group VIIB of the periodic table are called the *halogens*–the salt producers. They are fluorine, chlorine, bromine, iodine and astatine. The first two are gases, bromine is a liquid, and iodine is a crystalline solid. Astatine is a man-made radioactive element first produced in 1940. It is the odd-man-out of the group, showing little resemblance to the other halogens and as yet remains merely a chemical curiosity.

Fluorine

The first member of the halogen group and the most reactive is fluorine. Although it had long been known combined in its salts the fluorides, the free element was not prepared until 1886 by the French chemist Henri Moissan. He made it by electrolysing an ice-cold solution of potassium hydrogen fluoride in hydrogen fluoride.

Fluorine is an extremely reactive element. It attacks most metals and glass. With hydrogen, it reacts explosively to form hydrogen fluoride. This compound is generally made by warming calcium fluoride with concentrated sulphuric acid in a lead or platinum dish:

$$CaF_2 + H_2SO_4 \rightarrow CaSO_4 + H_2F_2.$$

Notice that the formula of hydrogen fluoride gas is written as H_2F_2. This is because its molecules are *dimeric*–they are each made up of two molecules of HF. Below 19·5°C, the gas condenses to a liquid. Like fluorine, hydrogen fluoride is corrosive. It is used for etching designs on glass, and must be stored in rubber or plastic bottles. It dissolves in water to give a solution of hydrofluoric acid, which gives rise to a series of poisonous salts called fluorides. In very low concentrations (about one part per million), fluorides may be added to drinking water. Fluoridized water is said to reduce tooth decay.

The volatile liquids used in most refrigerators are called freons. These are organic compounds containing fluorine. The simplest ones can be regarded as methane, CH_4, with some of its hydrogens replaced by fluorine–as in fluoro-

Chlorine is highly poisonous and was used as a war gas during World War I.

form, CHF_3. Freons are also used as the propellant gases in aerosol sprays. Unlike most inorganic compounds of fluorine, they are not poisonous. Polymers of organic fluorine compounds, such as polytetrafluoroethylene (PTFE), are heat-resisting plastics used as electrical insulators and for lining non-stick frying pans.

Chlorine

Chlorine was discovered nearly 200 years ago by the Swedish chemist Karl Scheele. It is a greenish-yellow choking gas, used as a war gas at the Battle of the Somme in 1915. It is made by the action of electrolysis or oxidizing agent on chlorides. In the laboratory, chlorine is made from hydrochloric acid and manganese dioxide or potassium permanganate:

$$MnO_2 + 4HCl \rightarrow MnCl_2 + 2H_2O + Cl_2.$$

Commercially, chlorine is made by electrolysis. It generally crops up as a by-product in the electrolytic manufacture of other substances from chlorides – for example, in making caustic soda (sodium hydroxide) from brine (sea water):

$$2NaCl + 2H_2O \rightarrow 2NaOH + H_2 + Cl_2.$$

To prevent the chlorine evolved from reacting with the sodium hydroxide, specially designed cells are used to prevent the products from mixing (see page 84). Chlorine is also produced as a by-product in the electrolytic extraction of the metals calcium and magnesium from their chlorides.

Chlorine is highly reactive, though not so reactive as

fluorine. It combines explosively with hydrogen in the presence of sunlight to form hydrogen chloride. It is a powerful oxidizing agent and is used for killing germs in drinking water and swimming baths. Chlorine dissolves slightly in water, with which it reacts to form a mixture of hydrochloric and hypochlorous acids:

$$Cl_2 + H_2O \rightleftharpoons HCl + HOCl.$$

Hypochlorous acid and its salts (the hypochlorites) are used as strong bleach and disinfectant because they are powerful oxidizing agents:

$$X + HOCl \rightarrow XO + HCl.$$

Chlorine reacts with cold dilute solutions of caustic alkalis to form a mixture of chloride and hypochlorite:

$$2KOH + Cl_2 \rightarrow KCl + KOCl + H_2O.$$

This process was developed in France in the 1780s where the resulting solution was sold as a bleach called *Eau de Javelle*. It is still sold in many countries as a lavatory cleanser and for sterilizing babies' bottles.

With lime (calcium oxide, CaO), chlorine reacts to produce a white compound known as bleaching powder. It corresponds to a mixture of calcium chloride, $CaCl_2$, and calcium hypochlorite, $Ca(OCl)_2$, and is assigned the loose formula $CaOCl_2$. In the presence of dilute acid, bleaching powder (or chloride of lime) releases chlorine gas, used for bleaching cloth. Factories for making bleaching powder were built in Britain beginning in 1799 and were essential to the nineteenth century textile industry. With hot strong alkalis, chlorine forms salts called chlorates, which are used as weed killers and in making matches:

$$6KOH + 3Cl_2 \rightarrow 5KCl + KClO_3 + 3H_2O.$$

(Top) Hydrogen chloride is made in the laboratory by treating sodium chloride with hot concentrated sulphuric acid. It dissolves in water to form hydrochloric acid.
(Bottom) The commonest salt of hydrochloric acid is sodium chloride (common salt). It occurs as underground deposits of rock salt. Salt can be extracted from sea water by evaporation in salt pans.

Hydrochloric acid

This acid is made by dissolving hydrogen chloride gas in water. The gas is made commercially by burning a stream of hydrogen in chlorine. In the laboratory, it is made by the action of hot concentrated sulphuric acid on a chloride:

$$H_2SO_4 + NaCl \rightarrow NaHSO_4 + HCl.$$

This reaction accounts for the old name for hydrochloric acid which was 'spirits of salt'. Hydrogen chloride has a pungent acrid smell and fumes in moist air. It can be detected by the clouds of 'smoke' formed in damp air,

such as the air breathed out from the lungs. It gives an acid reaction with damp litmus paper.

Many metals dissolve in hydrochloric acid to form the corresponding metal chloride and hydrogen:

$$2HCl + Zn \rightarrow ZnCl_2 + H_2.$$

A solution of zinc chloride made this way, known as 'killed spirits of salt' is used as a flux in soldering. The acid is used for 'pickling' iron and steel to remove rust.

Salts of hydrochloric acid are called chlorides. Common salt, sodium chloride, occurs in sea water and together with potassium chloride in extensive underground deposits in Cheshire, Germany, and Siberia. Silver chloride is insoluble and provides a useful and sensitive test for chlorides. The solution to be tested is mixed with silver nitrate solution and a little dilute nitric acid. The formation of a precipitate of white silver chloride reveals the presence of a chloride:

$$AgNO_3 + KCl \rightarrow KNO_3 + AgCl\downarrow.$$

Ammonium chloride (old name sal ammoniac) can be made from ammonium hydroxide and hydrochloric acid. It is also formed as dense clouds of white smoke by direct union between ammonia gas and hydrogen chloride:

$$NH_3 + HCl \rightarrow NH_4Cl.$$

It is used in Leclanché cells and dry batteries (see page 29). Aluminium chloride is used as a catalyst in organic chemistry, such as in the 'cracking' of high molecular weight petroleum hydrocarbons to convert them into petrol and kerosene.

Bromine

Bromine is one of the two elements that are liquids at ordinary temperatures (the other is mercury). It can be made by displacing it from one of its salts (bromides) with chlorine. In the commercial process, chlorine gas is pumped into sea water, which contains small amounts of sodium bromide:

$$2NaBr + Cl_2 \rightarrow 2NaCl + Br_2.$$

It is made in the laboratory by adding hot concentrated sulphuric acid to a mixture of potassium bromide and manganese dioxide. Hydrogen bromide is first produced and is oxidized to bromine:

$$KBr + H_2SO_4 \rightarrow KHSO_4 + HBr,$$

$$4HBr + MnO_2 \rightarrow MnBr_2 + 2H_2O + Br_2.$$

Hydrogen bromide is best made by the action of water on phosphorus tribromide. This is achieved by making a paste of red phosphorus with water and slowly dripping bromine onto it. It dissolves in water to form hydrobromic acid which gives rise to the salts called bromides. Potassium bromide is used in medicine as a sedative and silver bromide,

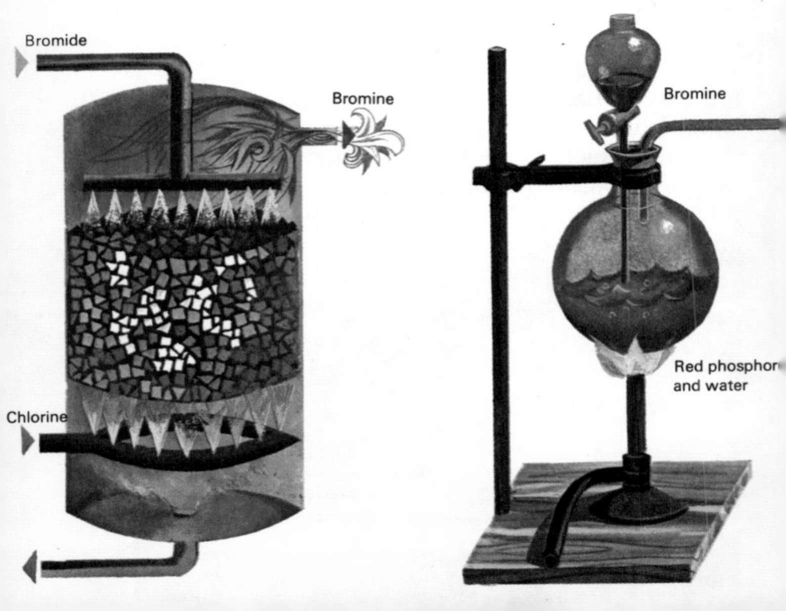

which is sensitive to light, is used for making photographic emulsions for films and printing papers. Bromine itself is important for making 'anti-knock' additives for petrol.

Iodine

Iodine is a black crystalline solid which vaporizes on heating to clouds of violet vapour. It is obtained commercially from sodium iodate which occurs as an impurity in deposits of Chile saltpetre (sodium nitrate). It was formerly made from kelp, the ashes of seaweeds which contain about one per cent iodine.

The laboratory preparation of iodine is similar to that of bromine. A mixture of potassium iodide and manganese dioxide is treated with hot concentrated sulphuric acid. The preparation of hydrogen iodide also resembles that of the bromine compound, but this time water is added to a mixture of iodine and red phosphorus. It dissolves in water to form hydriodic acid.

Iodine is essential in the human body for the functioning of the thyroid gland in the neck. People with insufficient iodine in their diet suffer from the disease called goitre in which the neck swells. In Derbyshire, where iodine is lacking in the soil, goitre was called Derbyshire neck.

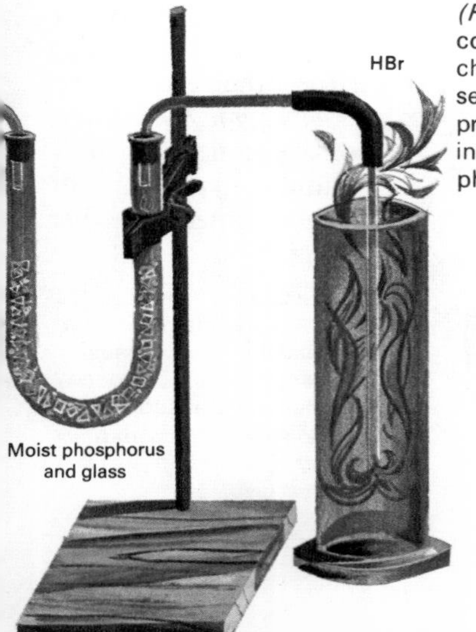

(Far left) Bromine is made commercially by the action of chlorine on the bromide salts in sea water. *(Left)* The laboratory preparation of hydrogen bromide involves treating moist red phosphorus with bromine.

CARBON

Quite early in the history of chemistry workers recognized that chemical compounds can be classified into two types: those derived from inanimate objects such as rocks and those derived from living things. In 1777, the Swedish chemist Torbern Bergman suggested that these two groups be called *inorganic* compounds and *organic* compounds. About fifty years later the classification was upset when Friedrich Wöhler succeeded in making the organic compound urea from the inorganic salt ammonium cyanate. As a result, chemists revised their definition so that now the chemistry of carbon compounds is called *organic chemistry*, whereas the study of all the other elements is called *inorganic chemistry*. Carbon is unique in forming compounds in which its atoms are joined together in chains.

Carbon exists as several allotropes—soot, diamond, and graphite (used in pencil leads and nuclear reactors). Diamond is the hardest natural substance and can be sawn only with a high-speed, diamond-impregnated disc.

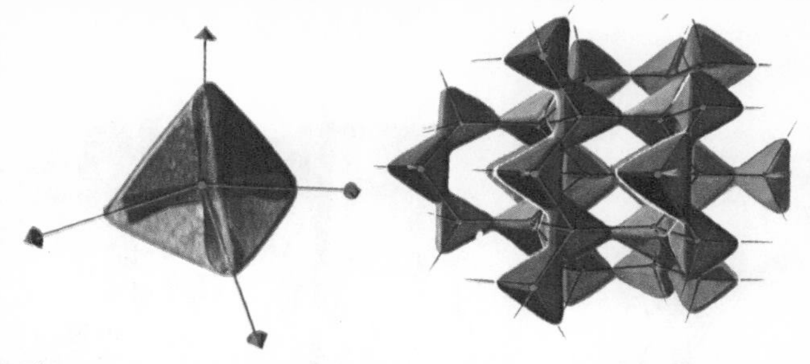

In diamond, the carbon atoms are rigidly bonded tetrahedrally in a crystal lattice, giving diamond its extreme hardness.

Diamond and graphite

This distinction must be modified because carbon, carbon dioxide, and carbonates are generally included in the study of inorganic chemistry. The gravels of river beds in Brazil and India contain diamonds, but most come from South African mines. These buried diamonds formed when pieces of other carbon allotropes (physical forms)–probably graphite or charcoal–were acted on by extremes of temperature and pressure in the Earth's crust. Artificial diamonds are made by reproducing these conditions in the laboratory.

In graphite, the carbon atoms are tightly bonded in sheets which are only loosely held together in layers. The layers are free to slide, giving graphite its lubricating properties.

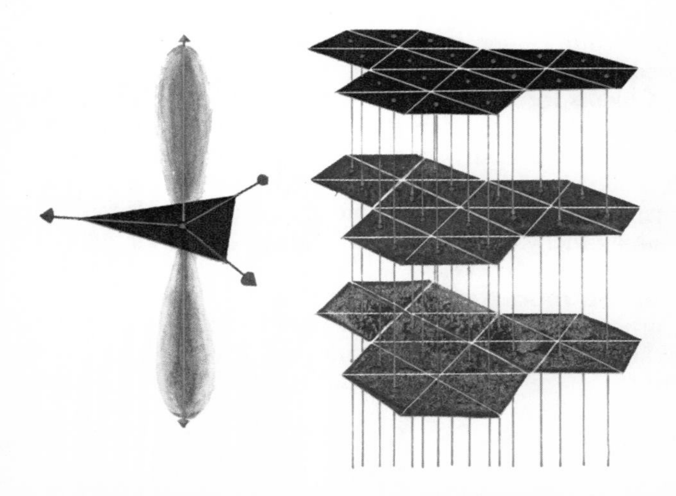

Coal and wood are rich in carbon; on controlled combustion they yield coke and charcoal, both almost pure forms of carbon.

Charcoal and coke

Charcoal and coke, as well as various forms of soot, are made up of tiny microcrystals of carbon. Heating wood in the absence of air gives the wood charcoal used by artists, and heating bone in a similar way gives bone charcoal, used for absorbing gases and impurities.

Coke is formed, together with various gases, when coal is heated in the absence of air. Coke is a common domestic fuel and is the form of carbon used in the chemical industry. It is added to blast furnaces for making steel where it has a chemical function – it is not merely a fuel. It is also used for making fuel gases such as producer gas and water gas.

Carbon monoxide

Carbon takes part in very few chemical reactions. But if it is heated in a stream of steam or if it burns in a limited supply of air or oxygen, carbon monoxide is formed.

The mixture with hydrogen produced by the first of these reactions is called water gas. When it is formed by the second reaction (passing air over red-hot coke), carbon monoxide is known as producer gas. It is also formed when a fuel such as coal or petrol is incompletely burnt, hence its presence in coal gas and motor car exhaust gases.

Carbon monoxide is formed in a coke fire with an inadequate air supply. The carbon (coke) burns to produce carbon dioxide which is reduced to carbon monoxide on passing up through the fire:

$$CO_2 + C \rightarrow 2CO.$$

In the form called 'after damp', it is also found in coal mines after an explosion of methane ('fire damp'). Carbon monoxide is made in the laboratory by treating oxalic or formic acid with hot concentrated sulphuric acid.

When carbon, in any of its forms, burns completely it produces carbon dioxide. In a limited supply of air or oxygen it yields carbon monoxide, a fuel gas used to power London buses during World War II.

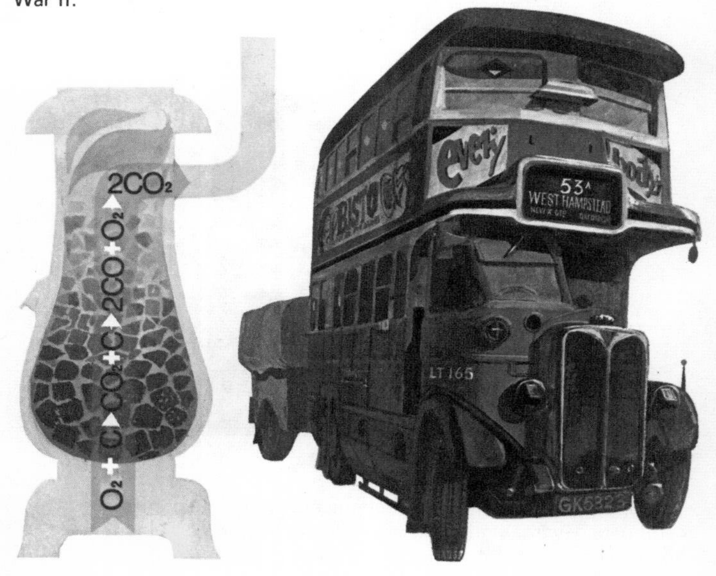

The White Cliffs of Dover are vast deposits of calcium carbonate, formed on the bed of an ancient ocean from the shells of billions of tiny sea creatures.

Carbon dioxide

When carbon burns in a plentiful supply of air or oxygen, it forms carbon dioxide:

$$C + O_2 \rightarrow CO_2.$$

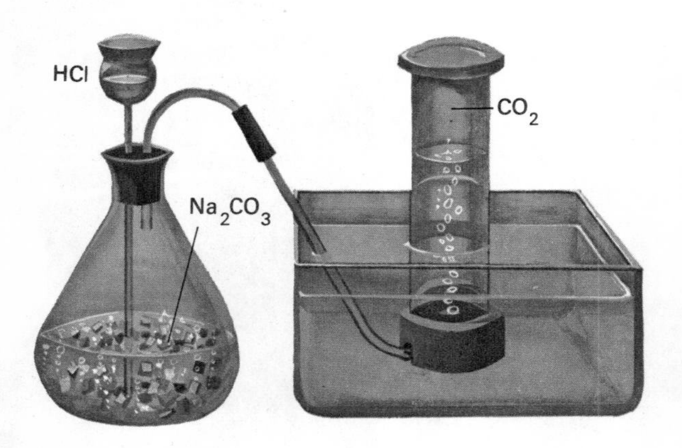

Carbon dioxide is generated by the action of any acid on any carbonate. Hydrochloric acid and calcium carbonate in the form of marble chips are used in the laboratory.

It is also generated when certain carbonates are heated, as in the manufacture of lime by roasting chalk (calcium carbonate):

$$CaCO_3 \rightarrow CaO + CO_2$$

or by the action of acid on a carbonate or bicarbonate:

$$Na_2CO_3 + 2HCl \rightarrow 2NaCl + CO_2 + H_2O.$$

Carbon dioxide dissolves in water, and under pressure gives soda water. A little of the carbon dioxide reacts chemically with the water to form carbonic acid, H_2CO_3, which gives soda water its refreshing, tangy taste.

At low temperatures, carbon dioxide solidifies. The solid formed is known as 'dry ice' because it turns straight back to the gas on warming without passing through a liquid stage as does ordinary ice. Some fire extinguishers contain carbon dioxide under pressure which forms clouds of tiny dry ice crystals when the valve is opened.

Carbonates

These are the salts of carbonic acid. The most important ones are sodium carbonate (washing soda) and calcium carbonate, which occurs in many forms. The chalk cliffs of Dover, marble and sea shells are all natural forms of calcium carbonate. Sodium carbonate, commonly used as an alkali for making soap and glass, is manufactured from salt and carbon dioxide by the Solvay process.

Hydrocarbons

These, the simplest organic chemicals, are compounds of hydrogen and carbon. They are of two types: those derived from fatty substances such as oil and petroleum, called *aliphatic* hydrocarbons, and those derived from coal, called *aromatic* hydrocarbons. There is an important structural difference between the two types. In aliphatic compounds, the carbon atoms link with each other to form chains of atoms. In aromatic compounds, they join together in sixes to form rings of atoms.

Aliphatic hydrocarbons

These compounds are divided into three main

Petroleum is the major source of aliphatic hydrocarbons.

groups, depending on their chemistry and structure. The first group begins with the compounds methane, ethane, propane, and butane, which have the formulae CH_4, C_2H_6, C_3H_8 and C_4H_{10}. There is a pattern in these formulae and the group, called the paraffins, forms a series of compounds of general formula C_nH_{2n+2}, where *n* is a whole number. Such a group is an *homologous series*. In the case of the paraffins, all the carbon bonds are joined to different carbon atoms or to hydrogen and so they are known as *saturated* compounds.

The first members of the paraffin series are gases, commonly used as fuels. Methane occurs in coal gas and in the rotting vegetation at the bottoms of ponds–hence its old name marsh gas. Propane and butane are sold in pressurized cylinders as the fuel used by campers and caravaners. The liquid paraffins include petrol, ordinary paraffin (or kerosene), diesel oil and lubricating oil. They occur in petroleum and are separated by distillation at an oil refinery. The lower members of the series are also made by 'cracking' the higher members with the aid of a catalyst.

The compounds of the next homologous series each have two of their carbon atoms joined by a double bond–they are *unsaturated* compounds. They are called *olefines* and the first members are ethylene, C_2H_4, and propylene, C_3H_6, illustrating the general formula C_nH_{2n}. Unlike the paraffins, olefines are chemically reactive because one of the carbon-carbon double bonds is easily broken. Hydrogen will add across the double bond of higher members such as ground nut oil and whale oil. This converts the liquids to solid paraffins, and is the basis of the manufacture of margarine and cooking fat.

The third homologous series of aliphatic hydrocarbons has a carbon-carbon triple bond and the general formula C_nH_{2n-2}. The series is named after its first and most important member acetylene, C_2H_2. Acetylene is a fuel gas used with oxygen to produce the oxyacetylene flame for welding. It is made by direct combination between the elements or by the action of water on calcium carbide:

$$CaC_2 + H_2O \rightarrow CaO + H_2C_2.$$

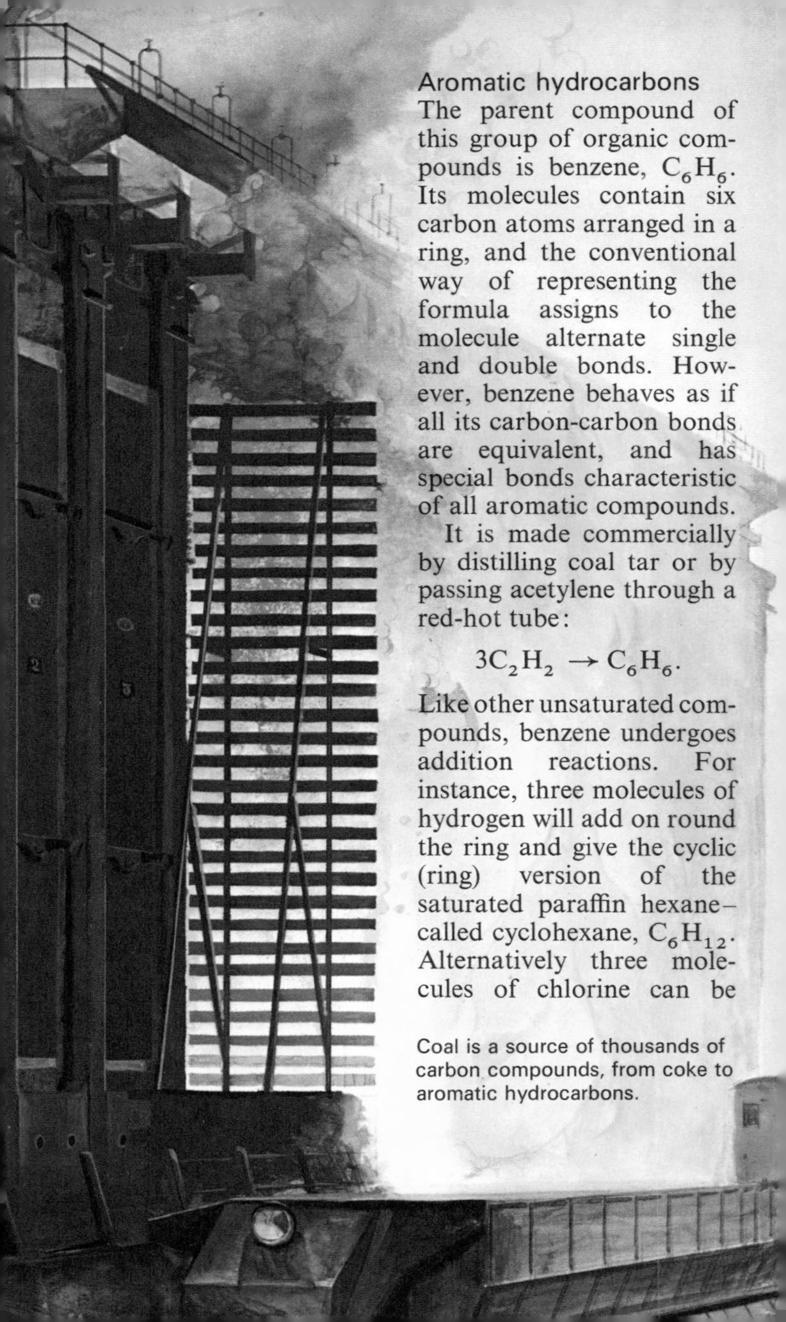

Aromatic hydrocarbons

The parent compound of this group of organic compounds is benzene, C_6H_6. Its molecules contain six carbon atoms arranged in a ring, and the conventional way of representing the formula assigns to the molecule alternate single and double bonds. However, benzene behaves as if all its carbon-carbon bonds are equivalent, and has special bonds characteristic of all aromatic compounds.

It is made commercially by distilling coal tar or by passing acetylene through a red-hot tube:

$$3C_2H_2 \rightarrow C_6H_6.$$

Like other unsaturated compounds, benzene undergoes addition reactions. For instance, three molecules of hydrogen will add on round the ring and give the cyclic (ring) version of the saturated paraffin hexane– called cyclohexane, C_6H_{12}. Alternatively three molecules of chlorine can be

Coal is a source of thousands of carbon compounds, from coke to aromatic hydrocarbons.

made to add on in a similar way to give the compound hexachlorocyclohexane which, under a legion of names (gammexane, hexane, benzene hexachloride, BHC, hexachlorophene), is used as an insecticide and bactericide.

More important are the *substitution* reactions of benzene – that is, reactions in which one or more of the hydrogen atoms are replaced by other atoms or groups. If an atom of chlorine is substituted for one of benzene's hydrogen atoms, the compound chlorobenzene is formed; substitution by a nitro group, NO_2, gives nitrobenzene. Some compounds of this type have special names which do not immediately indicate their relationship to benzene. For example, substitution by a methyl group, CH_3, gives *toluene* – not 'methyl benzene'. Similarly a hydroxyl group, OH, gives *phenol* and an amino group, NH_2, *aniline*.

There are six replaceable hydrogen atoms in one molecule of benzene and more than one of them can be substituted. Two chlorine substituents give dichlorobenzene, three give trichlorobenzene, and so on. With two substituents, there result three different possible compounds. The two chlorines in dichlorobenzene may substitute next-door to each other, next-door-but-one, or opposite each other. Chemists call them *ortho-*, *meta-* and *para-*, often abbreviated to *o-*, *m-* and *p-*.

An alternative naming scheme numbers the positions round the benzene ring 1 to 6. The first substituent enters at position one. For example, *m*-dichlorobenzene can also be called 1:3-dichlorobenzene. When a second substituent enters the ring of one of the specially named compounds mentioned earlier, the new compound is named in terms of it. For instance, when aniline is chlorinated in the ring the resulting compounds are called *chloroaniline*.

There are three ways in which a pair of substituents can enter the benzene ring, giving products designated ortho, meta and para.

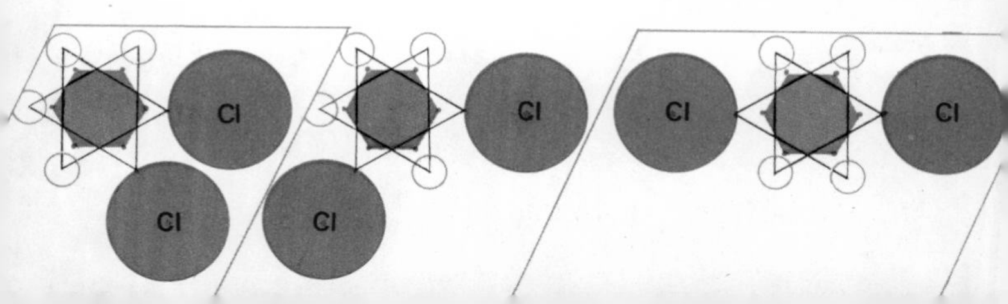

The commonest alcohol, ethanol, is the chief intoxicating component of beer, wines and spirits.

Alcohols

If one of the hydrogens of a paraffin is replaced by a hydroxyl group, OH, the resulting compound is called an *alcohol*. The simplest alcohol is *methanol*, CH_3OH, (formerly called methyl alcohol), which corresponds to methane, CH_4, with one of its hydrogens replaced by hydroxyl. The same substitution in ethane gives *ethanol* (ethyl alcohol), C_2H_5OH, and so on down the paraffin series.

Ethanol is probably the oldest known organic chemical.

Ethylene glycol, used as an antifreeze in car radiators, is an alcohol containing two hydroxyl groups.

It was discovered by alchemists about a thousand years ago and produced by distilling wine. It is the ordinary alcohol in spirits such as brandy and gin, and familiar to most people in the impure form sold as methylated spirits.

Ethanol is formed in beer and wines by fermentation, a process in which yeast breaks down sugars in fruit juices into ethanol and carbon dioxide. Beer froths and some wines 'sparkle' because they contain dissolved carbon dioxide. Methanol, once known as 'wood alcohol' can be obtained by distilling wood. It is extremely poisonous, small quantities causing blindness and death.

A compound with two hydroxyl groups is called a dihydric alcohol or *glycol*. For example, two hydroxyl substituents in ethane give ethylene glycol, $CH_2OH \cdot CH_2OH$. This oily liquid has a sweet taste (although it is poisonous) rather than the fiery taste of ethanol. It is used mixed with water as an antifreeze in motor car radiators.

The simplest trihydric alcohol, formed by substituting three hydroxyl groups into butane, is *glycerol* (or glycerine), $CH_2OH \cdot CHOH \cdot CH_2OH$. It is a non-poisonous, sweet-tasting liquid. The hydroxyl groups react with nitric acid to form the explosive glyceryl trinitrate, better known as nitroglycerine.

Phenol

Substitution of a hydroxyl group for one of the hydrogens in the benzene ring gives phenol, C_6H_5OH. Although structurally this compound resembles an alcohol, in its properties it behaves much more like an acid–hence its old name carbolic acid (from the Latin for coal). The hydrogen of the hydroxyl group may be replaced by a metal to form a salt called a phenate. Certain other substituents in phenol's benzene ring make the hydroxyl part even more acid. For instance, trinitrophenol, better known as picric acid, is quite a strong acid–and a powerful explosive. Phenol is extracted from coal tar. It may be made from chlorobenzene or aniline and is used as a disinfectant and as a starting material for many drugs and dyes.

Aldehydes and ketones

If hydrogen is removed from an alcohol, the resulting compound is called an *aldehyde*. Methanol gives formaldehyde, H·CHO, and ethanol gives acetaldehyde, $CH_3 \cdot CHO$. Aldehydes are named after their parent acids rather than the alcohols. The removal of hydrogen from an alcohol is generally facilitated by an oxidizing agent such as potassium dichromate:

$$C_2H_5OH + [O] \rightarrow$$

$$CH_3.CHO + H_2O.$$

Formaldehyde is a gas. Its solution in water is called formalin and is used for 'pickling' biological specimens and, with phenol, for making Bakelite-type plastics. Acetaldehyde is a liquid which polymerizes to the solid metaldehyde, used as a fuel in portable stoves. The higher aldehydes are also liquids with a fruity smell made use of in perfumes.

Formalin is a solution of formaldehyde in water, used for preserving biological specimens *(above)*.

The explosive hexogen is made from formaldehyde, ammonia and nitric acid *(left)*.

An unusual condensation reaction takes place between formaldehyde and ammonia. The resulting compound has the name hexamethylenetetramine. It is used in medicine under the name urotropine as a bladder disinfectant. With nitric acid it forms a trinitro compound called hexogen which is a high explosive used in torpedo warheads.

Aldehydes are formed by the oxidation of primary alcohols. Oxidation of a secondary alcohol gives a *ketone*. For example, oxidation of *iso*-butanol gives acetone, $CH_3{\cdot}CO{\cdot}CH_3$. Two alkyl groups astride the carbonyl group, CO, are here the same and so acetone is an example of a *simple* ketone. If they are different, the compound is known as a *mixed* ketone. The simplest mixed ketone is methyl ethyl ketone, $CH_3{\cdot}CO{\cdot}C_2H_5$.

The lower ketones are liquids. Acetone is an important solvent for cellulose and was formerly used in nail varnish remover. The higher ketones have fragrant, 'flowery' smells and are used in perfumes.

Aromatic analogues

Benzene and its relatives also form aldehydes and ketones. The simplest aldehyde is benzaldehyde, $C_6H_5{\cdot}CHO$, made by the oxidation of toluene. It is oily liquid that smells of bitter almonds. Other aromatic aldehydes include vanillin, which gives the flavour to ice cream and custard, and anisaldehyde, also used in sweets and perfumes. The simplest aromatic ketone is acetophenone, $C_6H_5{\cdot}CO{\cdot}CH_3$ (phenyl methyl ketone). Strictly this compound is a mixed aliphatic-aromatic ketone and the simplest purely aromatic one is benzophenone (diphenyl ketone), $C_6H_5{\cdot}CO{\cdot}C_6H_5$.

Portable campers' stoves burn metaldehyde, a solid polymer of acetaldehyde.

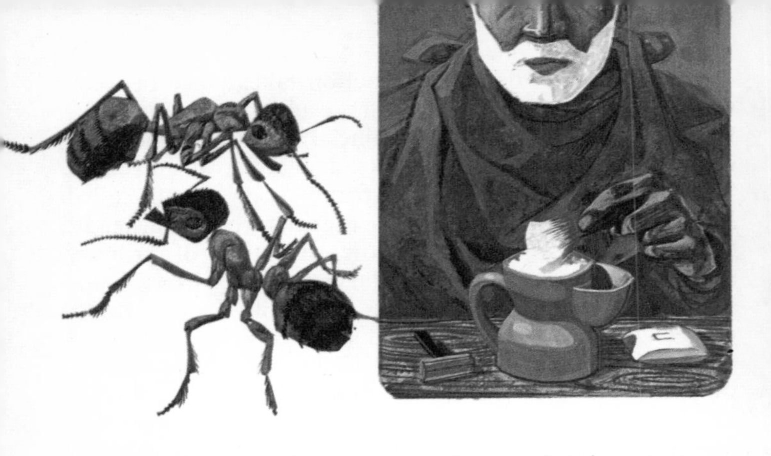

Formic acid is secreted by ants—scientific name *Formica*—as a means of defence *(left)*. Soaps are alkali metal salts of long chain fatty acids such as stearic acid *(right)*.

Acids and esters

If the oxidation of an alcohol is allowed to proceed beyond the aldehyde stage, a compound called a *carboxylic acid* is formed. For instance, the oxidation of alcohol gives acetic acid, $CH_3 \cdot COOH$, which is the process by which vinegar has been made for centuries.

The simplest carboxylic acid and first member of the series is formic acid, HCOOH. It is named after the biological name for the ant, *Formica,* which injects the acid into the wounds of its victims. It is made by the oxidation of methanol. The higher members of the series are often called fatty acids. They include palmitic and stearic acids, which occur as compounds with glycerol in oils and fats.

The hydrogen of the carboxyl group may be replaced by a metal to form a salt. Thus sodium acetate is formed by the reaction between sodium hydroxide and acetic acid.

Salt formation by carboxylic acids is another example of the general rule 'acid plus base gives salt plus water'. With alcohols, they undergo a similar reaction to form substances known as *esters;* acetic acid reacts with ethanol to form the ester ethyl acetate:

$$CH_3COOH + C_2H_5OH \rightarrow CH_3 \cdot COOC_2H_5 + H_2O.$$

Ether was one of the earliest anaesthetics.

Low molecular weight esters have fruity smells and are used in artificial flavourings. When an ester is treated with hot alkali, it splits up to re-form the alcohol and the corresponding salt of the acid. This process, virtually the opposite of esterification, is called *saponification.*

Most oils and fats are esters formed between glycerol (an alcohol) and high molecular weight carboxylic acids. Palm oil and olive oil contain glyceryl stearate. When these fats are treated with sodium hydroxide, they saponify to release glycerol and form salts such as sodium stearate. These salts are soaps (*sapo* is Latin for *soap*), and this reaction is the basis of soap manufacture.

Sodium stearate has the formula $C_{17}H_{35}COONa$. In warm water it is hydrolysed to form stearic acid and sodium hydroxide – that is why soap gives an alkaline reaction in solution. Calcium salts present in hard water react with the stearic acid to form an insoluble scum of calcium stearate.

Ether

When ethanol is heated with concentrated sulphuric acid, a molecule of water is knocked out from a pair of ethanol molecules to form diethyl ether:

$$2C_2H_5OH \rightarrow (C_2H_5)_2O + H_2O.$$

Diethyl ether (or simply 'ether' or, formerly, sulphuric ether) is used as a solvent, anaesthetic and fuel.

Chlorinated hydrocarbons

One or more hydrogen atoms in a paraffin may be substituted by chlorine atoms. Successive replacement of the four hydrogens of methane, CH_4, gives methyl chloride, CH_3Cl, methylene chloride, CH_2Cl_2, chloroform, $CHCl_3$, and carbon tetrachloride, CCl_4. A similar set of compounds may be based on ethane, propane and so on.

Methyl chloride is a volatile liquid used as a local anaesthetic (when allowed to evaporate from the skin it produces a sensation of intense coldness and numbness). It is made by esterifying methanol with hydrochloric acid:

$$CH_3OH + HCl \rightarrow CH_3Cl + H_2O.$$

It is one of a series of compounds called *alkyl halides*, with the general formula RX. R may be any alkyl group (methyl, ethyl, propyl, and so on) and X may be any halide (fluoride, chloride, bromide, or iodide). Alkyl halides have many applications in synthetic organic chemistry where they are used to introduce an alkyl group into other compounds. Toluene may be made from benzene and methyl chloride with an aluminium chloride catalyst (this type of reaction is known as the Friedel-Crafts synthesis):

$$C_6H_6 + CH_3Cl \rightarrow C_6H_5 \cdot CH_3 + HCl.$$

Methylene chloride, chloroform, and carbon tetrachloride are all volatile liquids used as solvents and anaesthetics. Methylene chloride is made by the partial reduction of chloroform:

$$CHCl_3 + H_2 \rightarrow CH_2Cl_2 + HCl.$$

Chloroform is made by treating ethanol with chlorine in alkaline solution. It is oxidized in air to form the poisonous substance phosgene, used as a war gas in World War I. For this reason, chloroform for anaesthetic use contains a small amount of ethanol to prevent phosgene accumulating. The bromine and iodine analogues of chloroform – bromoform and iodoform – are both used in medicine, iodoform being a yellow solid with a strong 'antiseptic' smell. Fluoroform, CHF_3, is used in refrigerators and aerosols.

Carbon tetrachloride is a good grease solvent used in the dry cleaning of clothes. Its advantage over other solvents (such as ether and ethanol) is that it is non-inflammable; it is even used in fire extinguishers, when it is effective against petrol fires.

There are similar chlorine derivatives of ethane. They are generally prepared by chlorination of acetylene, and include tetrachloroethane, $CHCl_2 \cdot CHCl_2$ (used as a solvent for cellulose compounds), trichloroethylene, $CHCl : CCl_2$ (used as an industrial solvent and, under the name Trilene, as an anaesthetic), and hexachloroethane, $CCl_3 \cdot CCl_3$ (which smells like and is used as a substitute for camphor). Among the higher molecular weight chlorine derivatives is *pp′*-dichlorodiphenyl(trichloromethyl) methane. Under the name DDT, this compound is used as an insecticide.

The paraffins are not the only class of aliphatic compounds that can have hydrogen atoms substituted by chlorine or another halogen.

Chlorination of carboxylic acids produces various chloro-acids, which are much stronger than the acids from which they are derived. Acetic acid, $CH_3 \cdot COOH$, forms successively monochloroacetic acid, $CH_2Cl \cdot COOH$, dichloracetic acid, $CHCl_2 \cdot COOH$, and trichloracetic acid, $CCl_3 \cdot COOH$.

Carbon tetrachloride is used in fire extinguishers and as a grease solvent in dry cleaning.

HCl
Na OH

Amines

Substitution of a hydrogen atom of a paraffin by an amino group, NH_2, gives rise to a series of compounds called *amines*. Replacement of a hydrogen of methane, CH_4, by an amino group gives *methylamine*, CH_3NH_2, made by treating an alcoholic solution of ammonia with methyl chloride:

$$NH_3 + CH_3Cl \rightarrow CH_3NH_2 + HCl.$$

Because methylamine has only one alkyl group, it is called a *primary* amine. But two alkyl groups can substitute in ammonia to form a *secondary* amine such as dimethylamine, $(CH_3)_2NH$, and three substituents give a *tertiary* amine such as trimethylamine, $(CH_3)_3N$. It is even possible to replace all the hydrogens of an ammonium salt to give what is called a *quaternary ammonium salt*, such as tetra-methylammonium chloride, $(CH_3)_4NCl$. Like ammonia, amines are bases.

Aniline

If an amino group is substituted for one of the hydrogens in the benzene ring, the resulting compound is aniline, $C_6H_5NH_2$. It is generally made by reducing nitrobenzene with zinc and hydrochloric acid:

$$C_6H_5NO_2 + 6[H] \rightarrow C_6H_5NH_2 + 2H_2O.$$

Aniline is a colourless liquid found in coal tar. It is the starting material for the important synthetic aniline dyes.

Amides

These compounds have the general formula $RCONH_2$. The lower members of the series are formamide, $HCONH_2$ (a gas), and acetamide, CH_3CONH_2 (a solid which smells of mice). They may be made by heating the ammonium salt of a carboxylic acid:

$$CH_3COONH_4 \rightarrow CH_3CONH_2 + H_2O.$$

(Left) In 1865 William Perkin accidentally made a mauve dye from aniline, which eventually gave rise to the whole synthetic dyestuffs industry.

Sulphur compounds

In inorganic chemistry, sulphur behaves very much like oxygen in many of its reactions. Similarly in organic chemistry, many aliphatic compounds containing oxygen have analogues containing sulphur instead. The simplest of these are the *thiols* (also called *mercaptans*), which are the sulphur analogues of the alcohols. Methyl mercaptan, CH_3SH, and ethyl mercaptan, C_2H_5SH, may be considered to be methanol, CH_3OH, and ethanol, C_2H_5OH, with sulphur instead of oxygen. They are made by treating potassium hydrogen sulphide with an alkyl halide:

$$C_2H_5Cl + KSH \rightarrow C_2H_5SH + KCl.$$

Mercaptans have a penetrating vile smell resembling rotten cabbage. They are added in small quantities to odourless fuel gases so that leaking gas can be detected by its smell.

The sulphur analogues of ethers are called *thioethers*. They are made by treating potassium sulphide with an alkyl halide:

$$2C_2H_5Cl + K_2S \rightarrow (C_2H_5)_2S + 2KI.$$

They are used in the rubber and plastics industries.

Sulphonic acids

These are made by treating benzene or its relations with hot concentrated sulphuric acid:

$$C_6H_6 + H_2SO_4 \rightarrow C_6H_5SO_3H + H_2O.$$

Sulphonic acids are much used in synthetic organic chemistry because of the ease with which the sulphonic acid group may be replaced by other groups. For example with molten sodium hydroxide, benzene sulphonic acid gives phenol:

$$C_6H_5SO_3H + 2NaOH \rightarrow C_6H_5OH + Na_2SO_3 + H_2O.$$

The alkali metal salts of sulphonic acids, called sulphonates, are soluble in water and are the main constituents of detergents. The molecules of these compounds generally have a long-chain aliphatic group substituted in the benzene ring opposite the sulphonate group. As a result, the molecules have a long, grease-soluble tail (the aliphatic part) and a water-soluble head (the sulphonate part). Shampoos and washing-up liquids contain similar detergent substances.

The sulphonate molecules of detergents have grease-soluble tails and water-soluble heads and so can remove dirt and grease from clothes.

Organometallic compounds

Sulphur is not the only 'inorganic' element that will form organic compounds. Many metals and metalloids will too, including arsenic, antimony, bismuth, tin, lead and magnesium. One class of these so-called organometallic compounds common to most of the metals are called *alkyls*. They are formed between metal atoms and alkyl radicals and correspond to the hydride of the metal with the hydrogens replaced by alkyls.

Lead alkyls are poisonous, stable liquids. They have been well studied and one with four different alkyl groups has been prepared (methyl ethyl propyl butyl lead). Tetraethyl lead is added as an anti-knock agent to petrol and is made by the action of ethyl chloride on lead amalgam (an alloy with mercury):

$$3Pb + 4C_2H_5Cl \rightarrow Pb(C_2H_5)_4 + 2PbCl_2.$$

Free radicals

Tetraethyl lead works as an anti-knock agent because the heat in a motor car engine breaks it down into lead and free ethyl radicals:

$$Pb(C_2H_5)_4 \rightarrow Pb + 4C_2H_5\cdot$$

Free radicals are extremely reactive (in this case they combine with any volatile constituents in the petrol and prevent pre-ignition), and are used to trigger off many important chemical processes. Many plastics are made by polymerizing simple organic chemicals such as ethylene (which forms polyethylene, or polythene) and such polymerization reactions are sparked off by free radicals (see page 139).

Tetraethyl lead is used as an anti-knock ingredient in petrol, but is being banned in some countries to lessen atmospheric pollution.

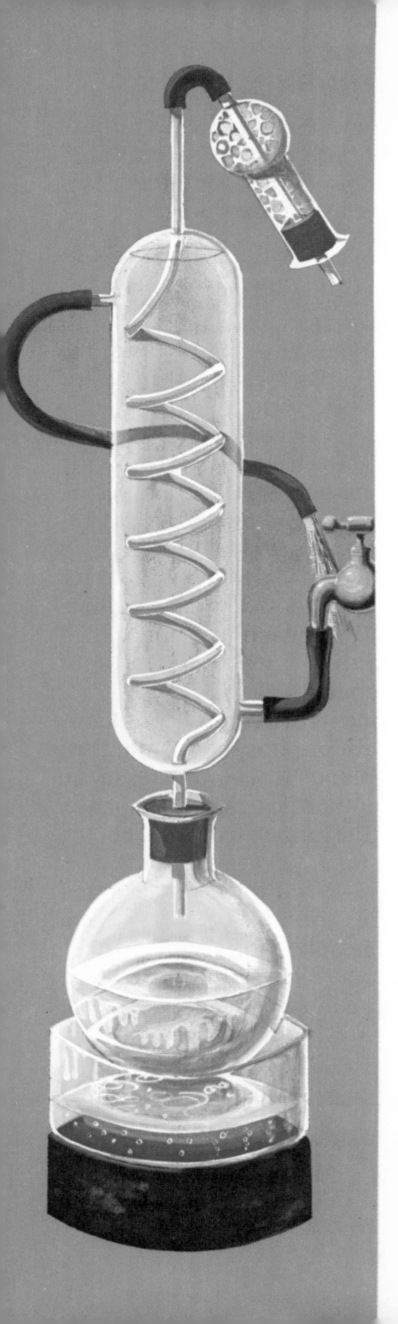

Grignard reagents

Compounds of magnesium with an alkyl group and a halogen, called alkylmagnesium salts or Grignard reagents, are extremely important in synthetic organic chemistry. They are made by reacting an alkyl halide with magnesium in ether:

$$CH_3Br + Mg \rightarrow CH_3MgBr.$$

If the reaction is slow, it may be speeded up by adding a small crystal of iodine. Grignard reagents are very reactive. With water they give a paraffin (in this case ethane):

$$C_2H_5MgCl + H_2O \rightarrow C_2H_6 + Mg(OH)Cl;$$

with carbon dioxide and water a carboxylic acid:

$$CH_3MgBr + CO_2 + H_2O \rightarrow CH_3COOH + Mg(OH)Cl$$

and with sulphur and water they give a mercaptan:

$$C_2H_5 + S + H_2O \rightarrow C_2H_5SH + Mg(OH)Cl.$$

One of the most useful synthetic techniques in organic chemistry employs Grignard reagents, compounds of magnesium and an alkyl halide.

PHOSPHORUS AND SULPHUR

Phosphorus and sulphur show some resemblances to carbon; they are both non-metals, they both exist as several allotropes, and they both form acidic oxides. But these two elements are much more reactive than is carbon.

Phosphorus

There are three allotropes of phosphorus, called white, red and black. White (or yellow) phosphorus is a waxy solid. It is extremely inflammable and catches fire spontaneously in air to form dense white clouds of phosphorus pentoxide. For this reason it is generally kept under water. It is made by condensing phosphorus vapour, produced by heating rock phosphate with sand and coke in an electric furnace. White phosphorus is extremely poisonous and is used in rat poison.

Red phosphorus is non-poisonous and much less inflammable. It is made by heating white phosphorus to more than 260°C in an inert atmosphere. It is used in the striking surface on safety match boxes. The third allotrope, black phosphorus, is made by heating yellow phosphorus under high pressure. Its atoms are arranged in layers (like those of graphite) and, like graphite, it consists of flaky crystals.

Phosphorus is an essential element for living things. Plants take it in from the soil in the form of phosphates, and men and animals take it in by eating plants. It accumulates in the bones, which are composed mainly of calcium phosphate, although some is excreted in the urine–from which the element was first isolated more than 300 years ago.

Phosphorus forms two oxides. When burnt in a limited supply of air it gives phosphorus trioxide P_2O_3, and in a plentiful supply of air it forms phosphorus pentoxide P_2O_5. Both oxides react readily with water (hence they are used as drying agents) to form more than a dozen different oxyacids. The most important of these are phosphorous acid (the trioxide and cold water):

$$P_2O_3 + 3H_2O \rightarrow 2H_3PO_3,$$

orthophosphoric acid (the pentoxide and warm water):

$$P_2O_5 + 3H_2O \rightarrow 2H_3PO_4,$$

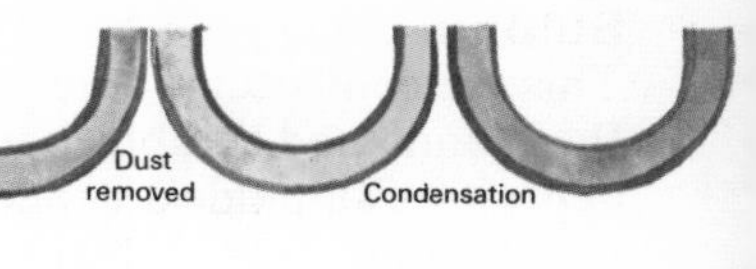

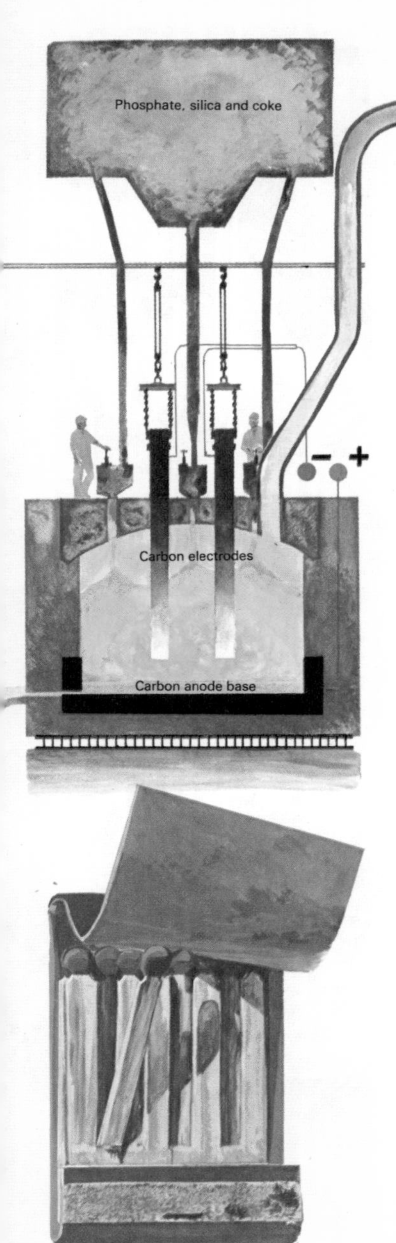

and metaphosphoric acid (the pentoxide and cold water):

$$P_2O_5 + H_2O \rightarrow 2HPO_3.$$

The salts of these acids are called phosphites and phosphates. The most important phosphate is calcium phosphate $Ca_3(PO_4)_2$. It occurs in the mineral rock phosphate or apatite and is converted into a much more soluble form for use as a fertilizer called 'superphosphate' by treatment with sulphuric acid.

Phosphorus also forms two chlorides (the trichloride PCl_3 and the pentachloride PCl_5) and a gaseous hydride, PH_3, called phosphine. This gas, made by treating white phosphorus with hot sodium hydroxide solution, is inflammable and has a fishy smell.

Phosphorus is extracted by reducing phosphate rock with coke and sand in an electric furnace. Red phosphorus is used in making matches.

Sulphur

This element occurs free in the ground and combined in deposits of sulphide ores and of sulphate salts. Native sulphur is found in volcanic regions, such as Sicily, where it is mined. The crude sulphur is melted, using more crude sulphur sometimes mixed with coal as a fuel, and the molten sulphur runs off into moulds. Sulphur also occurs deep in the ground in the United States associated with deposits of petroleum. It is extracted by the ingenious Frasch process, named after its inventor Herman Frasch. Wells are drilled into the sulphur beds and a pipe consisting of three concentric tubes pushed down. Water is superheated to 180°C under pressure (so that it does not boil) and pumped down the outside tube. The hot water melts the sulphur. Compressed air is pumped down the centre tube, and a frothy mixture of air, water and

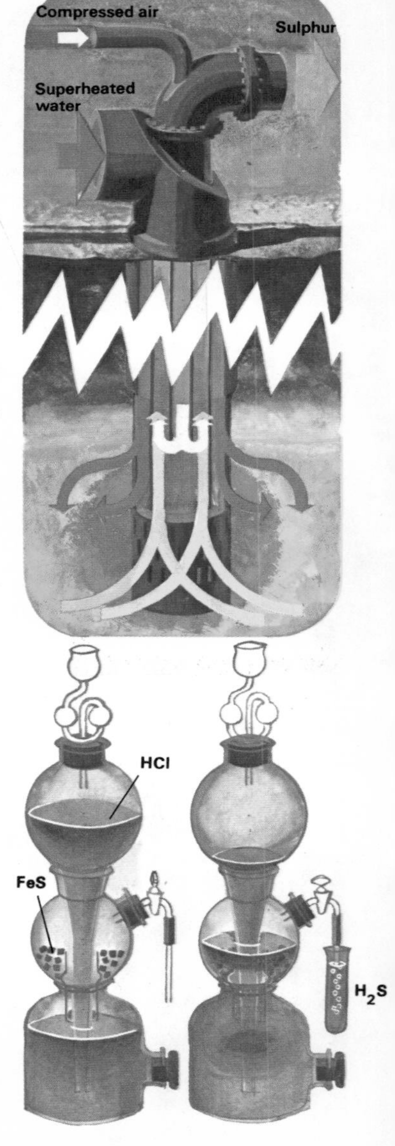

Underground deposits of sulphur are melted and extracted in the Frasch process *(above)*. A continuous laboratory supply of hydrogen sulphide gas, much used in chemical analysis, is provided by a Kipp's apparatus *(right)*.

molten sulphur comes up the third tube to be run off into huge settling tanks. Four-fifths of the world's output of sulphur is produced in Louisiana and Texas by the Frasch process.

As molten sulphur cools below 115°C, it solidifies to needle-shaped crystals of *monoclinic* sulphur. Below 95·5°C it changes to the second allotrope, called *rhombic* sulphur, which has squarer crystals. Sulphur may be purified by boiling it and condensing the vapour. The crystals form flower-like patterns on the walls of the condensation chamber and so this powder form of the element is called flowers of sulphur. If molten sulphur is poured into cold water, it takes on a brownish rubbery form called plastic sulphur, which soon goes brittle and crystalline on further cooling.

Most of the sulphur produced is used for making sulphuric acid. The rest is converted into fungicides, fertilizers, insecticides, explosives, drugs, dyes, paints, plastics, oils, and detergents.

Chemically sulphur is a reactive element. It burns in air or oxygen to form the gas sulphur dioxide:

$$S + O_2 \rightarrow SO_2,$$

and in the presence of a catalyst will combine with more oxygen to form sulphur trioxide, SO_3 (this reaction is the basis of the manufacture of sulphuric acid). In the laboratory, sulphur dioxide is prepared by the action of an acid on a sulphite:

$$Na_2SO_3 + 2HCl \rightarrow 2NaCl + H_2O + SO_2.$$

Hydrogen sulphide

Sulphur combines with hydrogen to form the gas hydrogen sulphide, H_2S (formerly called sulphuretted hydrogen), which smells of bad eggs and is mainly responsible for the schoolboy's nickname of 'stinks' for chemistry. Again the laboratory preparation is different, the gas generally being prepared by the action of an acid on a sulphide:

$$FeS + 2HCl \rightarrow FeCl_2 + H_2S.$$

Hydrogen sulphide is an important reagent in analytical chemistry, and most laboratories have a device called a Kipp's apparatus to give a constant supply of the gas 'on tap'.

Sulphuric acid

There are two main processes for manufacturing sulphuric acid: the older lead-chamber process, and the contact process which has largely replaced it. Both processes convert sulphur dioxide into sulphur trioxide, which dissolves in water to form sulphuric acid. In the lead-chamber process, nitrogen dioxide does the oxidation:

$$SO_2 + NO_2 \rightarrow SO_3 + NO.$$

Excess air converts the nitric oxide formed back to nitrogen dioxide. All the reactions take place at once in a large, lead-lined chamber and the process yields an acid of about 60 per cent concentration.

The contact process gives a stronger and purer acid. In the presence of a catalyst such as platinum or vanadium pentoxide, oxygen reacts directly with sulphur dioxide at temperatures of up to 800°C. The sulphur trioxide formed is dissolved in concentrated sulphuric acid to form 'fuming' sulphuric acid or oleum, which may be diluted to give any required concentration.

In the obsolete lead chamber process for making sulphuric acid, sulphur dioxide is oxidized by nitrogen dioxide.

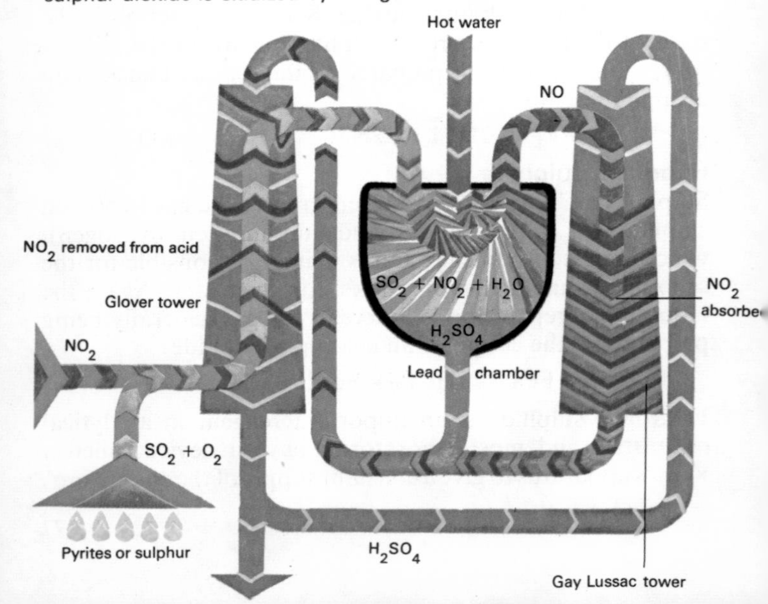

Practically every industry uses some sulphuric acid. Nearly half the world's production goes into making fertilizers such as superphosphate and ammonium sulphate. Organic compounds made with the aid of sulphuric acid include sulpha drugs, dyes and detergents. The acid is also used in oil refineries, making paints, and as a weed-killer. Some is used in making paper, plastics, and explosives. In the metal industry, it is used for 'pickling' iron and steel to remove rust and scale.

Chemically sulphuric acid behaves as a typical acid. The dilute acid reacts with metals, oxides, hydroxides, or carbonates to form salts called sulphates. Concentrated sulphuric acid displaces other acids from their salts (hence its use in making hydrogen chloride) and is used as a drying agent because of its strong affinity for water. The hot acid is an oxidizing agent, and so instead of evolving hydrogen with a metal (as does the dilute acid), it generates sulphur dioxide.

In the contact process, sulphur dioxide reacts with oxygen in the presence of a catalyst to form sulphur trioxide, which is converted into sulphuric acid.

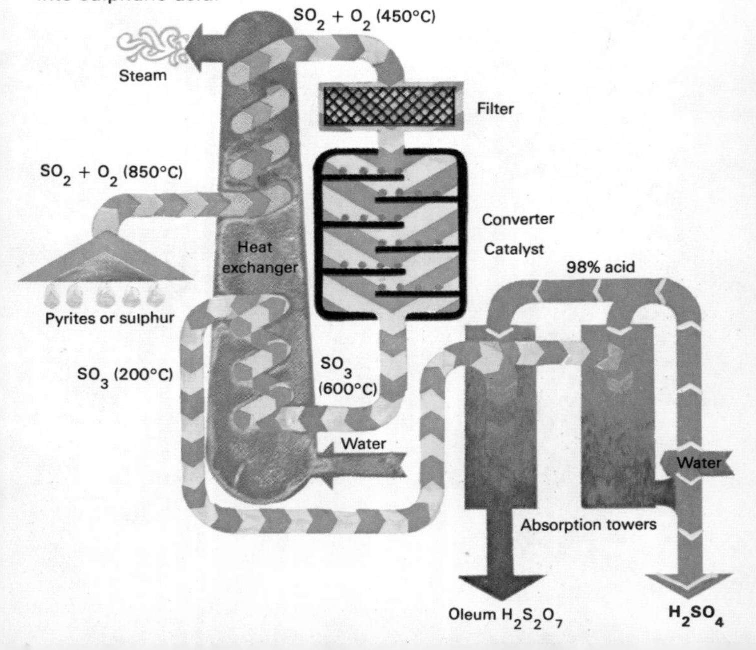

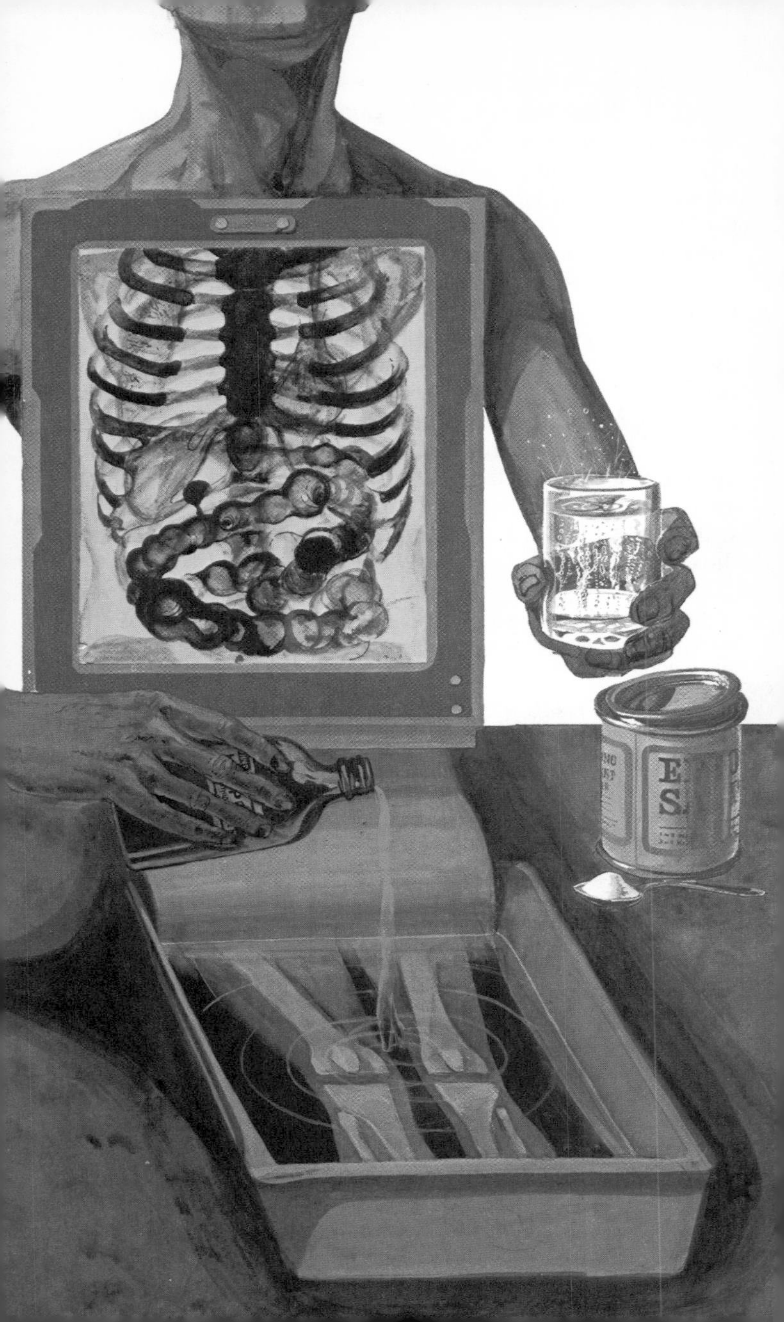

Sulphates

The salts of sulphuric acid include many minerals and other compounds of commercial value. Epsom salts (magnesium sulphate) are used as a laxative, as is Glauber's salt (sodium sulphate) which is more extensively used in making paper from wood pulp. Calcium sulphate occurs as the mineral gypsum, used for making plaster. On gentle heating it yields Plaster of Paris, a form of calcium sulphate with less associated water. Barium sulphate is a white paint pigment and the main constituent of the 'barium meal' drunk to make the stomach and intestines opaque to X-rays so that these organs can be photographed.

Ammonium sulphate is a nitrogenous fertilizer used both on the land and in the lakes of fish farms. It is a by-product of coal-gas works or may be made from synthetic ammonia and calcium sulphate. Aluminium sulphate, made from sulphuric acid and bauxite (aluminium oxide), is used as a mordant in dyeing. With the sulphate of an alkali metal, it forms an unusual series of double salts called alums. Potash alum or the substance known simply as 'alum' is the commonest of these, although other trivalent metals may replace aluminium in alums to give such compounds as ferric alum and chrome alum.

Sulphurous acid

Like phosphorus, sulphur also forms many oxy-acids. The most important of these next to sulphuric acid is sulphurous acid, H_2SO_3, which may be made by dissolving sulphur dioxide in water. Its salts, the sulphites, yield sulphur dioxide when treated with an acid. Thiosulphuric acid is best known as its sodium salt, sodium thiosulphate or photographer's 'hypo'. It is made by boiling sodium sulphite solution with sulphur:

$$Na_2SO_3 + S \rightarrow Na_2S_2O_3.$$

Useful sulphates include barium sulphate, which is given in the form of a barium meal to make the stomach and intestines opaque to X-rays, and magnesium sulphate, better known as the laxative Epsom salts. Sodium thiosulphate is the photographer's hypo, used for fixing a developed film or photographic paper.

THE ALKALI METALS

The metals of Group IA of the periodic table all dissolve in water, with a violence that varies from vigorous to explosive, to form hydrogen and an alkali. For example, sodium forms sodium hydroxide:

$$2Na + 2H_2O \rightarrow 2NaOH + H_2.$$

For this reason, these elements are called the alkali metals. They are lithium, sodium, potassium, rubidium, caesium and the radioactive element francium.

Sodium

This element occurs as sodium chloride in sea water and salt deposits in such places as Cheshire and Siberia. Its other common compound is sodium hydroxide (caustic soda) from which it was first extracted in 1807 by the British chemist Sir Humphrey Davy. His method, electrolysis, is

When Benvenuto Cellini made this ornate salt cellar, the metal sodium—found in common salt—had not been discovered.

still used commercially today although the cheaper chloride is used instead of the hydroxide.

There are very few uses for sodium as a metal. It is so reactive that it has to be kept under a layer of kerosene, and its strong affinity for water makes it a useful drying agent in organic chemistry. Some nuclear reactors have liquid sodium as the fluid in their heat interchanger.

When sodium or any of its compounds burn, they do so with a characteristic bright yellow flame. The same colour is produced by an electrical discharge passed through a tube containing sodium ions, and this is the basis of the familiar sodium lamps used for street lighting. The other alkali metals also produce characteristically coloured flames, and it was by these that the German chemist Robert Bunsen first detected the elements rubidium and caesium. He named them after Latin words meaning *dark red* and *sky-blue*.

Sodium is extracted by the electrolysis of molten sodium chloride. It is used in yellow street lamps which are lit by an electrical discharge through sodium vapour.

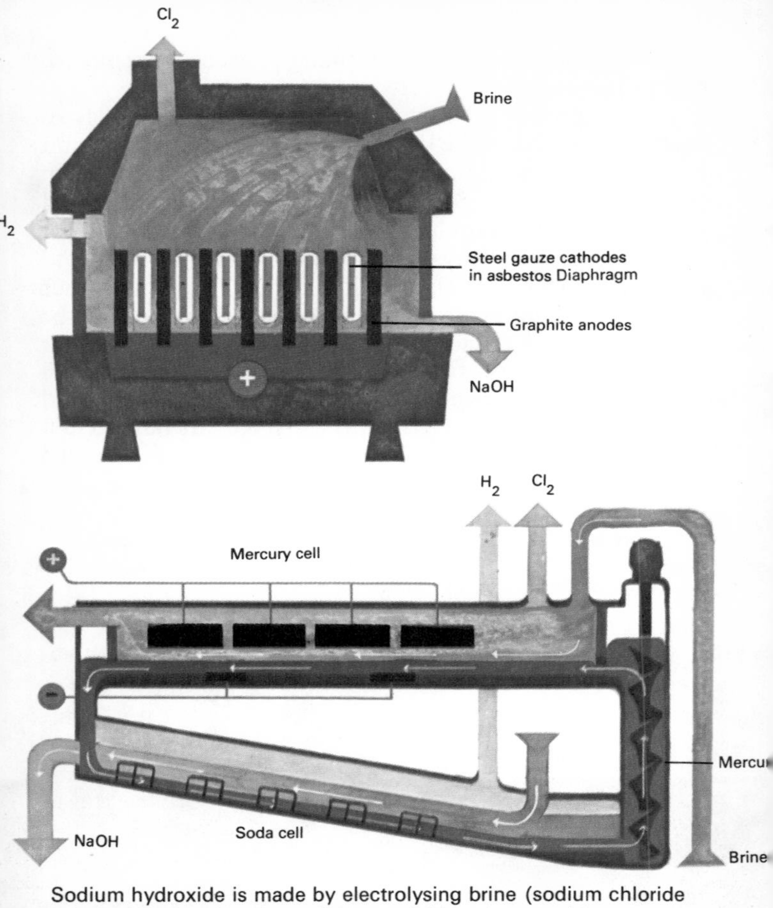

Sodium hydroxide is made by electrolysing brine (sodium chloride solution) and letting the sodium metal formed react with water either straight away, as in the Hooker cell *(top)*, or after being combined with mercury, as in the Castner-Kelner cell *(centre)*.

Sodium hydroxide

This is made from sodium carbonate solution and calcium hydroxide by double decomposition, or by the electrolysis of sodium chloride solution. The primary products of electrolysis are sodium and chlorine (which have to be kept apart): sodium is allowed to react with water. There are two ways of doing this commercially. In the porous diaphragm cell, hot brine is electrolysed in a porous pot with a carbon anode inside it and a steel mesh cathode outside. In the Castner-Kelner cell, the anodes are also carbon but the cathode consists of a pool of mercury. Sodium dissolves in the mercury, which is run off to react with water elsewhere.

Sodium hydroxide is used in making artificial silk and soap. It reacts with wood-pulp to leave cellulose and with oils and fats to leave soap.

Potassium

This alkali metal was also discovered by Davy in 1807, who electrolysed molten potassium hydroxide. Today the metal is extracted from the molten chloride using a carbon anode and an iron cathode. It occurs in the minerals feldspar and carnallite. Feldspar is an aluminosilicate and consequently not a readily available source of potassium. But carnallite, $KCl.MgCl_2.6H_2O$, occurs as extensive underground deposits at Stassfurt in Germany and is a useful source of potassium and magnesium. The waters of the Dead Sea are rich in potassium chloride.

Potassium metal closely resembles sodium. It is a soft, silvery metal which tarnishes rapidly in air and reacts violently with water to form hydrogen and potassium hydroxide. The heat of this reaction is often sufficient to ignite the hydrogen which, because of the presence of potassium, burns with a purple flame. This flame is characteristic of potassium and used to identify it in flame tests, although it is easily masked by the bright yellow flame of sodium.

Potassium compounds resemble those of sodium except that they are generally more soluble. Potassium nitrate (saltpetre) and potassium chlorate are important oxidizing agents. The former is a constituent of gunpowder and the latter is used in solid rocket fuels and in explosives. The oxidizing action of potassium chlorate also makes it a useful weed killer and a treatment for mouth ulcers. Potassium carbonate occurs in vegetable ash, hence its common name potash and the Latin name *kalium* (like the Arabic *al kali*, meaning *potash*) for the element.

THE ALKALINE EARTHS

The alkaline earth metals, like the alkali metals, are too reactive to occur in the free state, but they are not so reactive as the alkali metals. They get their name from the fact that their oxides, once called 'earths', give an alkaline reaction. For example calcium oxide (lime or quicklime) dissolves slightly in water to give an alkaline solution of calcium hydroxide (slaked lime). The other alkaline earth metals are magnesium, strontium, barium, and radium. Together with beryllium, they make up Group IIA of the periodic table. Beryllium more closely resembles its diagonal neighbour aluminium (Group III) than it does the other members of Group II.

Calcium

The most important member of the group, this metal occurs widely in the Earth's crust and is the third most abundant metal, next to aluminium and iron. Chalk, limestone, calcite, marble, sea shells, egg shells, and coral are all forms of calcium carbonate. That is why all these substances 'fizz' in acid to generate carbon dioxide and gives basis to the tradition that pearls–also a form of calcium carbonate–

Lime, cement and plaster all contain compounds of calcium.

dissolve in wine (if it is sour enough). Other calcium minerals are gypsum (calcium sulphate) and apatite (calcium phosphate). Gypsum is used for making plaster and blackboard chalk.

The action of strong heating converts calcium carbonate into calcium oxide. Traditionally, this is done in lime kilns when limestone is roasted to make lime:

$$CaCO_3 \rightarrow CaO + CO_2.$$

When heated strongly, calcium oxide incandesces; the bright white light it gives off was the original limelight used in theatres. Water reacts vigorously with lime to form calcium hydroxide or slaked lime:

$$CaO + H_2O \rightarrow Ca(OH)_2.$$

A dilute solution of calcium hydroxide, called lime water, is used as a test for carbon dioxide with which it forms an insoluble precipitate of calcium carbonate.

Lime and slaked lime are important chemicals to many industries. Cement is made by heating a mixture of limestone and clay, and glass by heating a mixture of lime, sand and soda ash (sodium carbonate). Lime is roasted with coke to make calcium carbide, the starting material for acetylene. Slaked lime mixed with sand forms mortar for laying bricks. It is also used for softening water and for neutralizing acidic soils. With chlorine it forms bleaching powder, once so important to Britain's textile industry.

Calcium metal is extracted by the same method used in its discovery by Sir Humphrey Davy in 1808: the electrolysis of the molten chloride. Calcium reacts fairly gently with cold water to form hydrogen and calcium hydroxide. Calcium chloride is deliquescent and is used as a drying agent in the laboratory.

Magnesium

This alkaline earth metal occurs as its chloride in sea water and in the mineral carnallite, and as its sulphate in Epsom salts. Dolomite is a mixture of calcium carbonate and magnesium carbonate, and magnesite is magnesium carbonate. Most magnesium is obtained from sea water by the

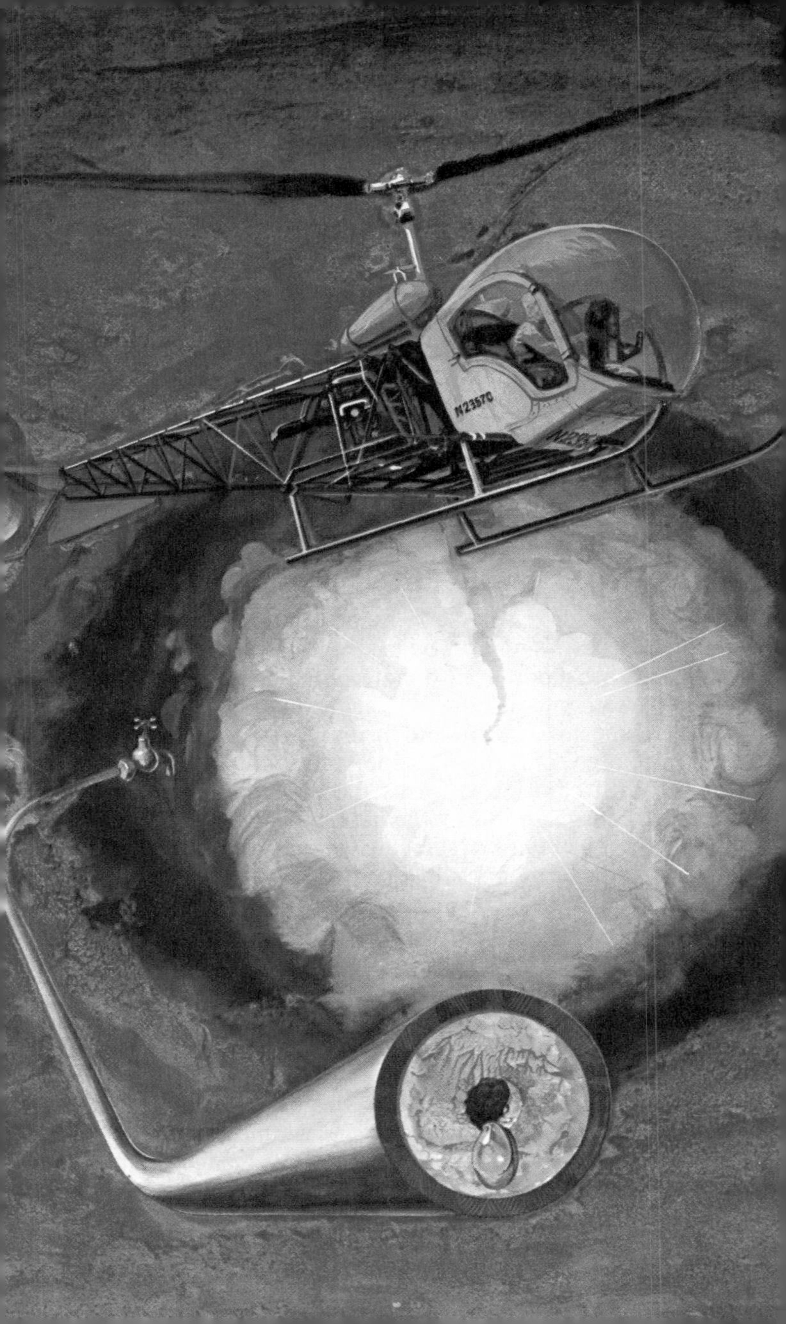
N2357C

electrolysis of the fused chloride. The pure metal tarnishes rapidly in air and reacts with steam to form hydrogen and magnesium hydroxide:

$$Mg + 2H_2O \rightarrow Mg(OH)_2 + H_2.$$

Magnesium hydroxide is generally made by treating a solution of a magnesium salt with sodium hydroxide. Its suspension in water, known as milk of magnesia, is used to treat stomach acidity. On heating, magnesium hydroxide gives magnesium oxide, a highly refractory substance used to line furnaces. Magnesium burns with a brilliant white light; once used by photographers, it is today reserved for flares and fireworks. Its alloys are extremely strong and light and much used in the aerospace industry. Magnesium is also used for cathodic protection of iron and steel structures; a piece of the metal connected to an iron pipe corrodes in preference to the iron.

Hard and soft water

Hard water, that is water which will not easily give a lather with soap, contains dissolved salts of calcium and magnesium. Temporary hardness consists of bicarbonates, formed when rain water containing dissolved carbon dioxide filters through limestone or chalk rocks:

$$CaCO_3 + H_2CO_3 \rightarrow Ca(HCO_3)_2.$$

It is called temporary because it is precipitated on boiling, and the calcium carbonate so formed sticks as hard 'fur' or scale on the insides of hot water pipes and kettles. Permanent hardness consists of the sulphates and chlorides of calcium and magnesium. These salts react with soap (sodium stearate) to form an insoluble scum (calcium stearate).

To prevent the waste of soap or formation of scale, or to get purer water for chemical processes, natural water is softened. There are various ways of doing this. Temporary hardness is removed by boiling or by the addition of slaked

Magnesium alloys are much used in making aircraft, and the pure metal burns with a blinding flash in magnesium flares. Deposits of natural magnesium salts from tap water contribute to the 'fur' which can block hot-water pipes.

lime. The addition of washing soda (sodium carbonate) or calgon (a complex phosphate) gets rid of both temporary and permanent hardness by precipitating the calcium and magnesium as insoluble salts. Ion exchange water softeners contain a natural mineral such as permutite or a synthetic resin which has the property of exchanging harmless sodium ions for the offending calcium and magnesium ions. Most deionizing agents can be regenerated by pouring strong salt solution through them to displace the 'exchanged' calcium and magnesium ions and replace them with a new stock of sodium ions.

Strontium, barium and radium

There are few uses for strontium compounds. They produce an intense red flame in the flame test and so are used in flares and fireworks. A radioactive isotope of strontium, strontium-90, is produced in nuclear explosions. If it reaches the Earth in fallout it is absorbed into the bones of men and animals. It remains radioactive for a long time, taking twenty-six years to lose half its radioactivity, and so its accumulation in the bones of children can be very dangerous.

Barium compounds burn with a green flame and are also

Strontium salts give a red flame to fireworks, those of barium a green flame.

used in fireworks. Soluble barium salts are intensely poisonous. Barium sulphate is insoluble (its formation from barium chloride solution being a standard test for sulphates) and a testament to its insolubility is its use in medicine where it is swallowed to make the stomach opaque to X-rays. Barium sulphate is also used as a white paint pigment. The peroxide, BaO_2, is used for making hydrogen peroxide, which it forms with any acid. In practice sulphuric acid is used because the barium sulphate also formed can easily be filtered off:

$$BaO_2 + H_2SO_4 \rightarrow H_2O_2 + BaSO_4.$$

Radium is an intensely radioactive element discovered in the mineral pitchblende in 1898 by Pierre and Marie Curie. Its discovery gave a big stimulus to the study of radioactivity and hence theories about the structure of the atom. Madame Curie herself used the radioactivity from radium for taking the first 'X-ray' photographs. Today the element is used in the treatment of cancer. Radium decays with a half-life of 1620 years into the inert gas radon.

The radioactive metal radium was discovered by the Curies. At first used for making X-ray photographs, it stimulated research into other radioactive elements and nuclear fission.

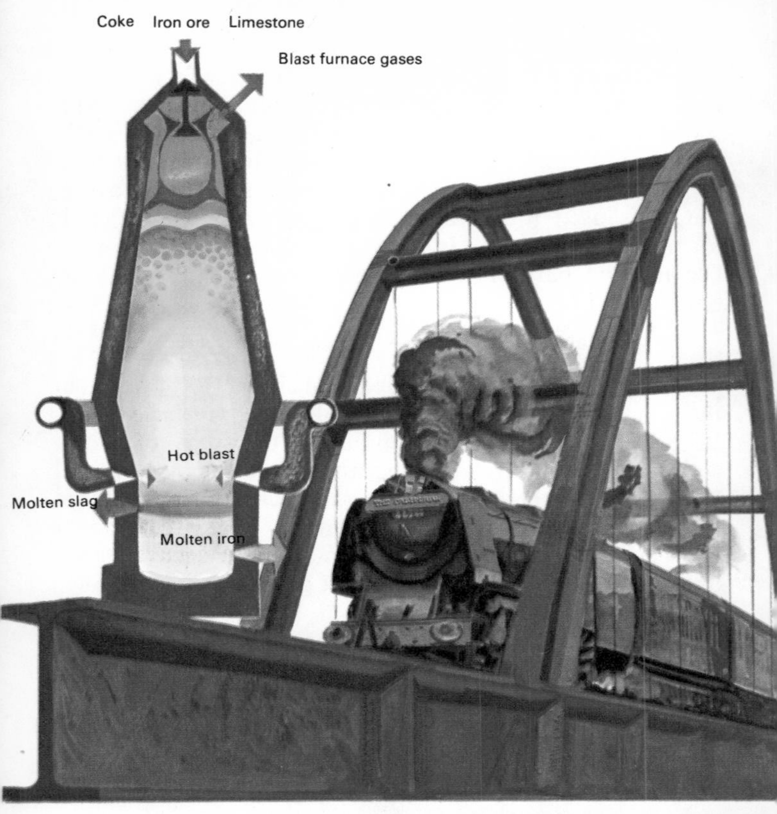
Coke
Iron ore
Limestone
Blast furnace gases
Hot blast
Molten slag
Molten iron

Iron

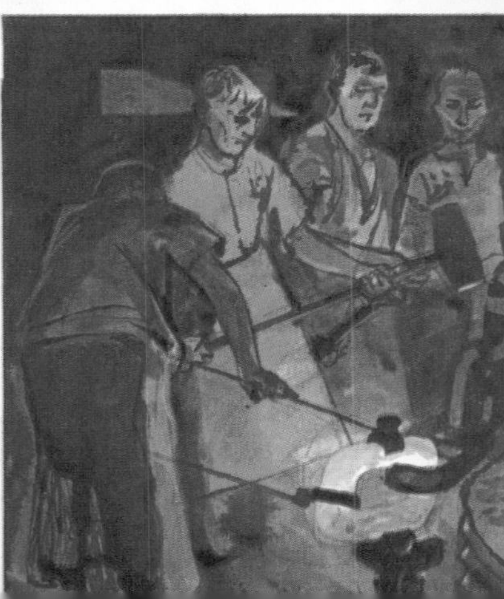

THE METALS

All but 25 of the 104 known elements are metals. The chemistry of two important groups has been discussed in the previous two sections, and most of the others will be described here.

Iron

Iron is a reactive metal and so is rarely found free in nature (except in meteorites). It is extracted from its ores, of which haematite, Fe_2O_3, and magnetite, Fe_3O_4, are the most common. The process takes place in a blast furnace, which is charged with a mixture of iron ore, coke and limestone. Air is blown into the bottom of the furnace and carbon monoxide reduces the iron oxide in the ore to metallic iron:

$$Fe_2O_3 + 3CO \rightarrow 2Fe + 3CO_2.$$

The limestone combines with impurities, which are mainly silica, to form a 'slag' of calcium silicate. When the hearth of the furnace is tapped, the iron runs out into comb-like moulds called 'pigs' to form pig iron. Pig iron is very impure and is used as a starting material for making cast iron, wrought iron, or steel.

Steel

Steel is the general name for alloys of iron. Ordinary steel contains a small percentage (less than 0·5 per cent) of carbon, but there are many special alloys such as tool steels, spring steels and stainless steels which also contain other metals. Steel may be made in a Bessemer converter or in an open-hearth or an electric furnace. The Bessemer process is quick, producing about 30 tons of steel in 15 minutes. In the open-hearth process, which takes about a day, a much larger quantity of steel is produced.

Iron is extracted from its ores in a blast furnace; it is the principal ingredient of steel, the most important metal to an industrialized nation. Before the invention of processes to make steel cheaply and in large quantities, many large metal objects were made from wrought iron, such as this anchor chain *(right)*.

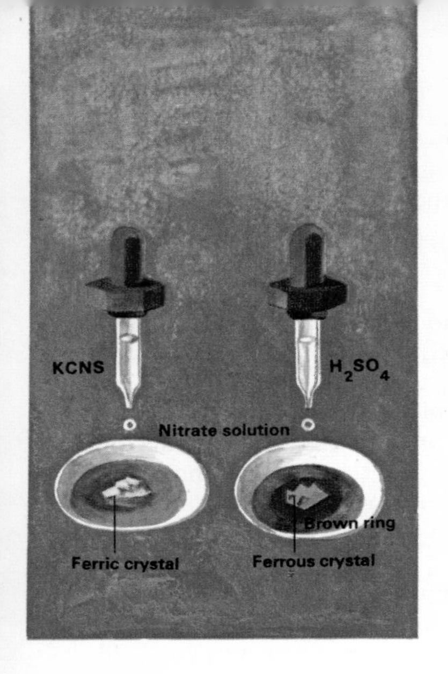

Two simple chemical tests distinguish between ferric and ferrous iron: ferric salts produce a blood-red colour with potassium thiocyanate solution, and ferrous salts give a brown coloration in the presence of a nitrate and concentrated sulphuric acid.

Chemistry of iron

Iron is a reactive metal. It has a variable valency and gives rise to two series of compounds, *ferrous* (divalent) and *ferric* (trivalent). For instance, iron dissolves in dilute sulphuric acid to form ferrous sulphate:

$$Fe + H_2SO_4 \rightarrow FeSO_4 + H_2,$$

but with hydrochloric acid it forms ferric chloride:

$$2Fe + 6HCl \rightarrow 2FeCl_3 + 3H_2.$$

Ferric chloride is an oxidizing agent; reducing agents such as hydrogen (zinc and hydrochloric acid) convert it to ferrous chloride. Oxidizing agents, even atmospheric oxygen, convert ferrous salts to ferric salts. There are two very sensitive tests for the two forms of iron. With potassium thiocyanate, ferric salts produce a bright blood-red colour;

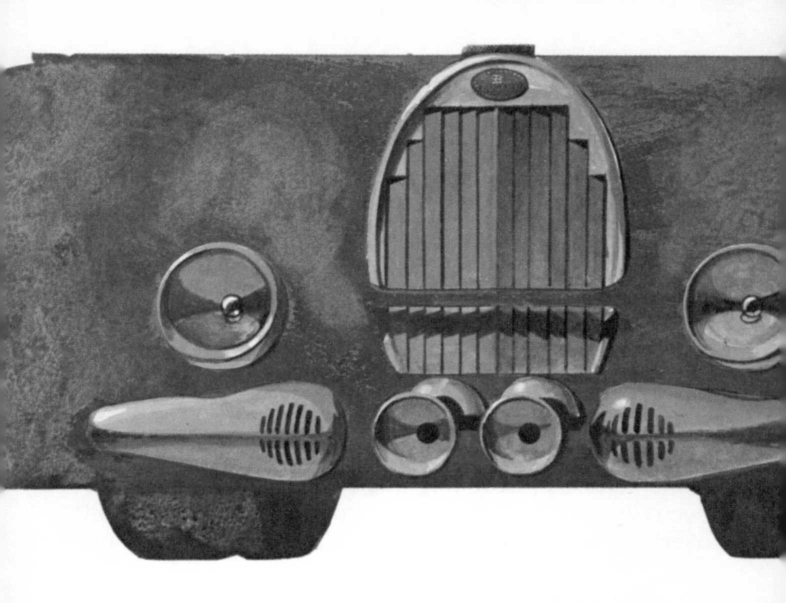

Iron, in the form of steel, is the chief metal used in making cars. Its chemical neighbour nickel is used in stainless steel and as a first coating in chromium plating.

with potassium ferricyanide, ferrous salts produce prussian blue. Ferrous sulphate is itself used as a reagent in the 'brown ring' test for nitrates.

Nickel

Nickel is a near neighbour of iron in Group VIII of the periodic table and resembles it in many ways. Chemically it is less reactive than iron and is often used in the form of electroplate to protect steel from corrosion. Alloyed with iron it gives stainless steel, and with copper it gives alloys known variously as German silver, nickel silver and cupro-nickel. An alloy with chromium, called nichrome, is used for making the elements of electric fires. Metallic nickel is used as a catalyst, as for example in the hydrogenation of vegetable oils to make margarine. It occurs mainly as the sulphide (sometimes mixed with copper and iron), especially in Canada which is the world's biggest producer.

Aluminium

The second most important metal (and the most abundant in the Earth's crust) is aluminium. Like iron it is strong and has a chemical valence of +3. But it is unlike iron in almost every other respect: it is light, it does not easily corrode, and it is difficult to extract from its ores.

Aluminium occurs abundantly as complex alumino-silicates in clay and in slate, but there is no process for extracting it from these sources. The chief ore is bauxite – aluminium oxide, Al_2O_3 – which is mined in tropical South America and Africa. The metal is extracted by an electrolytic process, and so the ore is shipped to countries such as Sweden and the United States where there is a plentiful supply of cheap hydroelectricity.

Aluminium is a good conductor of heat and so is used to make kettles and cooking utensils. Its good electrical conductivity and low density and make it ideal for high-voltage overhead wires. It does not corrode so it is used for making

Aluminium occurs in bauxite, from which it is extracted by electrolysis. China clay or kaolin, used for making porcelain, is also a compound of aluminium.

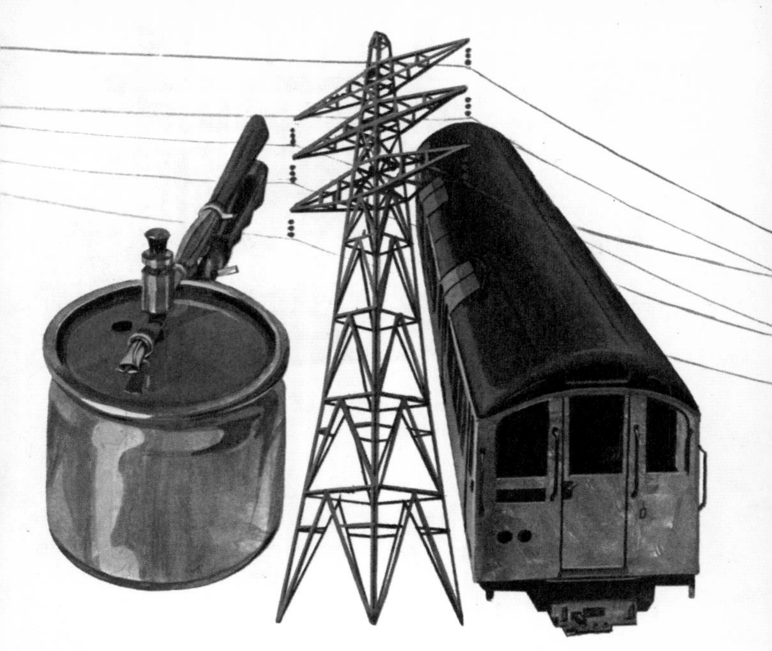

Aluminium is a good conductor of heat and electricity, and is used for making cooking utensils and high-voltage electric cables. Since it does not rust, coachwork made from it does not need painting.

reaction vessels in the chemical and petroleum industries. It can be polished to make reflectors or rolled into thin sheets to make aluminium foil for insulation and wrapping.

Chemically pure aluminium is fairly reactive. It dissolves in acids or in alkalis to give hydrogen. Because of this duality, it is called *amphoteric*. Aluminium oxide is also amphoteric. It dissolves in acids to give aluminium salts, and in akalis to give salts called aluminates.

In practice aluminium resists corrosion because it is always covered with a thin film of oxide. This protective oxide coating may be deliberately built up in the electrolytic process called *anodizing*. Sapphires and rubies are naturally occurring forms of aluminium oxide coloured by metallic impurities. Synthetic gems can be made by growing suitably doped crystals of alumina.

When copper is extracted by normal smelting methods the impure product is refined by electrolysis.

Copper

Copper was found in Cyprus (which is named after the Greek word for the metal) more than 5000 years ago. Alloyed with tin it produces bronze, the metal which marked the end of the Stone Age. Today copper is extracted from ores such as copper pyrites (cuprous sulphide, often mixed with iron), which is mined in Africa, North America and Scandinavia. The ore is smelted by roasting in a stream of hot air, and the metal purified by electrolysis.

Copper is a good conductor of electricity and half of the world's output is used to make wire and electrical contacts. It is also a good conductor of heat and is used for making cooking utensils and pipes for plumbing. Alloys of copper include brass (copper and zinc), bronze (copper and tin), and nickel silver (copper and nickel), all of which are

Copper is a good conductor of electricity and so is used for making electric wires and cables and for the conductors of printed circuit boards.

stronger and tougher than pure copper.

Chemically copper is fairly unreactive. It dissolves in nitric acid to give oxides of nitrogen, in hot concentrated sulphuric acid to give sulphur dioxide, and in a solution of ferric chloride:

$$2FeCl_3 + Cu \rightarrow 2FeCl_2 + CuCl_2.$$

This latter reaction is important in industry, and ferric chloride is used to etch copper to make printed circuits.

Zinc

The chief ores of zinc are zinc blende (zinc sulphide) and calamine (zinc carbonate). These are roasted in air to make the oxide, which is reduced to the metal by roasting with coke in a retort.

Zinc is a silvery bright metal but, like aluminium, soon becomes coated with a thin protective layer of oxide. In this way it resists corrosion and is used for protecting other metals such as steel, which may be electroplated with zinc or *galvanized* (dipped into molten zinc).

Zinc does not corrode easily and steel objects are coated with zinc (galvanized) to prevent them from rusting.

Gold
Silver

Silver

This metal, once used only for ornaments, jewellery and coins, is now the basic raw material for the vast photographic chemicals industry. Silver occurs either native (as the free element) or as the sulphide, when it is generally mixed with lead. It is extracted from the sulphide ores by roasting and is a profitable by-product of lead smelting.

Silver for jewellery may be pure or alloyed with copper (stirling silver) to make it harder. In air, silver tarnishes owing to the formation of oxide or sulphide.

Photographic emulsions contain silver bromide and/or chloride. Silver iodide is also used to seed snow crystals in rain-making.

Gold

This metal was formerly used entirely for jewellery,

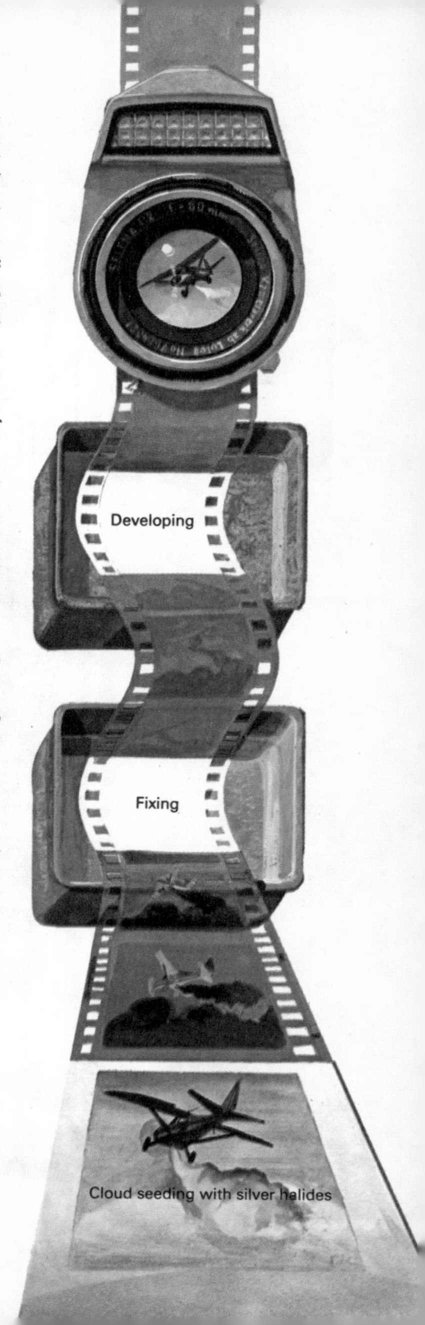

Gold and silver, precious metals that are easy to work, have been used for making jewellery and ornaments for more than 4 000 years *(left)*.
Silver halides are sensitive to light and are the basis of most photographic emulsions. The aircraft in the photograph is scattering silver iodide crystals into cloud to seed ice crystals and so make rain *(right)*.

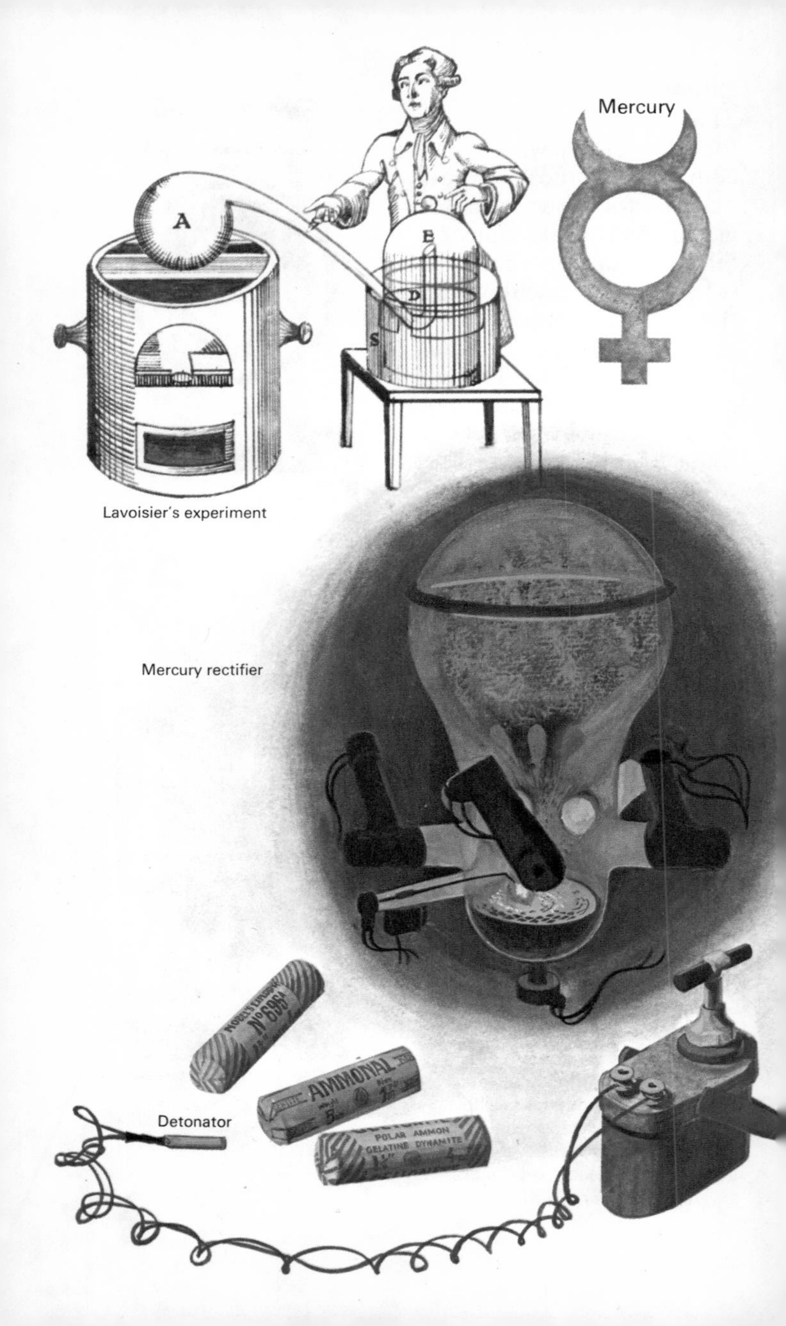
Mercury
A
B
D
S
Lavoisier's experiment
Mercury rectifier
Nº696
AMMONAL
POLAR AMMON
GELATINE DYNAMITE
Detonator

coins and bullion for national currency standards. In recent years, however, it has been employed by the electronics industry. Gold wires and films are used in making transistors and microminiature electronic circuits. Gold occurs native in river beds and mud (alluvial gold), from which it is extracted by 'panning'. Or it occurs as veins in rock and has to be mined.

Chemically gold is an extremely unreactive metal. It does not dissolve in any of the mineral acids, hence it was known as the noble metal. A mixture of nitric and hydrochloric acids, known as *aqua regia* (royal water), does dissolve gold.

Mercury

Mercury is that chemical curiosity a liquid metal. There is really nothing remarkable about mercury: all metals are liquid above their melting points, mercury just happens to melt at well below room temperature (at —39°C). It is a rare metal, occurring chiefly in Spain as its bright red sulphide ore called cinnabar (or vermilion). The metal is extracted simply by roasting the ore and condensing the mercury vapour. When heated in air, mercury forms its oxide which, on stronger heating, decomposes back to mercury and oxygen. This cycle of reactions was the basis of Lavoisier's famous experiments on oxygen and combustion.

Mercury is used in thermometers and barometers because it has a high coefficient of expansion and is liquid over a large range of temperatures. It is also used in mercury vapour lamps which give off an intense blue-green light rich in ultra-violet–hence their use in sun-tan lamps and photography. Many metals 'dissolve' in mercury to form alloys called amalgams. The most important of these contains silver and tin and is used in dentistry for filling teeth.

All mercury salts are poisonous, mercurous chloride

Mercury combines with oxygen on heating to form red mercury oxide which, on further heating, decomposes back to the metal and oxygen. These reactions were the basis of Lavoisier's classic experiments on combustion *(top)*. Today mercury is used in mercury rectifiers *(centre)* and, as mercury fulminate, in detonators for high explosives *(bottom)*.

(calomel) being used as an insecticide. Mercury fulminate is a highly sensitive explosive used for making percussion caps and detonators.

Cadmium

Cadmium is a rare metal resembling zinc and occurring with it in its ores blende and calamine. Cadmium metal absorbs neutrons and as a result is used to make the control rods for nuclear reactors. Cadmium sulphide, which varies in colour from bright yellow to orange, is used as a pigment in high-quality paints.

Tin

In about 800 BC, the Phoenicians sailed out of the Mediterranean Sea and northwards to the *Cassiterides*–the Tin Islands–that is to Britain, to the tin mines in what we call Cornwall. Today the chief ore of tin is still called cassiterite, although it is now mined mainly in Malaysia and Indonesia. It is stannic oxide and is smelted by being mixed with coal or coke and roasted in a reverberatory furnace.

Most of the world's tin production goes into making tinplate, used mainly for tin cans. Like zinc in galvanizing, tin

Tin is smelted in a reverberatory furnace, in which the heat is bounced back off a sloping ceiling. The metal is used mainly in coating steel strip for making 'tin' cans.

protects the steel from corrosion. Tin is also used to make tubes for holding paint or toothpaste, and is a component of many important alloys. Bronze is tin and copper, solder is tin and lead, and other tin alloys include typemetal and white metal or babbitt metal used for lining bearings.

Tin and its oxides are amphoteric. They dissolve in acids to form stannic (Sn^{4+}) or stannous (Sn^{2+}) salts and in alkalis to form stannates. Stannous chloride is a reducing agent, but stannic chloride is not salt-like at all. It is a volatile liquid, more like the covalent chlorides of non-metals.

Lead

Lead is another metal that was known to early civilizations. The Romans used it for casting statues and for making plumbing (the Latin word *plumbum* means lead). Today its use in plumbing has largely been replaced by copper and its other major use, as a protective sheath for electric cables, is being taken over by plastics. It is still used for making the electrodes in storage batteries (accumulators), typemetal, and shielding to protect people from gamma rays and X-rays.

The most widespread ore of lead is the sulphide, called galena. The metal is extracted by roasting the ore in a reverberatory furnace, or by converting it to the oxide and heating with coke in a blast furnace. Like so many chemically reactive metals, lead owes its non-corrodibility to the formation of an adherent surface coating of oxide.

Lead has several oxides. Lead monoxide or litharge, PbO, is used in making glass, lead dioxide, PbO_2, in accumulators, and red lead (triplumbic tetroxide, Pb_3O_4) in paint for protecting steel.

Rare earths

In Group III of the periodic table, the 15 elements from atomic number 57 to 71 are squeezed into one 'place'. They are all metals of very similar chemical properties known as the rare earth metals. More correctly, they are called the *lanthanides* after their first member lanthanum. The others are cerium, praseodymium, neodymium, promethium, samarium, europium, gadolinium, terbium, dysprosium, holmium, erbium, thulium, ytterbium and lutecium. The reason why all these elements are grouped together lies in the arrangement of electrons within their atoms.

Up to element 57, each successive element in the periodic table can be thought of as being formed by adding an electron (plus a proton and one or more neutrons) to the atom of the preceding element. In this way, the atomic number of

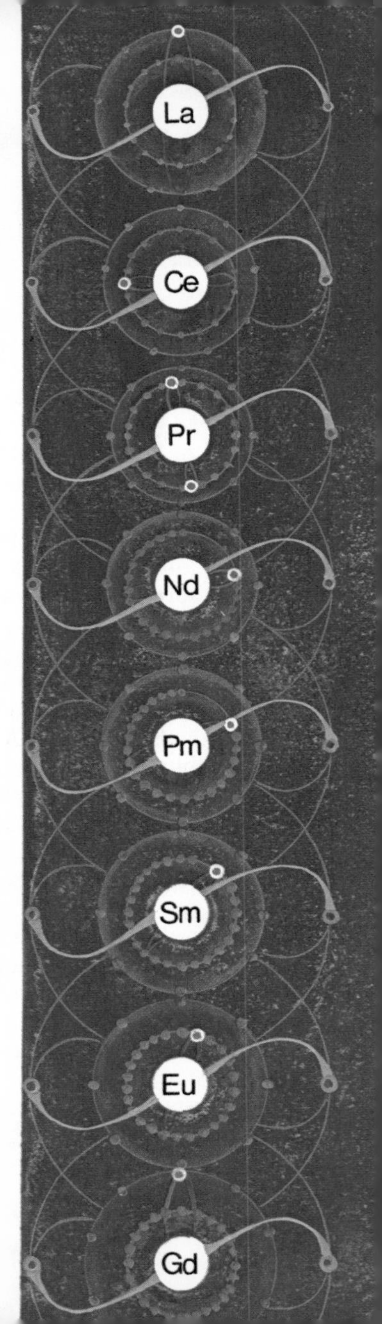

Each of the fourteen lanthanide elements has one electron more than its predecessor in the periodic table. The 'extra' electron generally enters an inner shell, leaving the configuration of the outer shells the same; as a result, the chemistry of the lanthanides is very similar.

each successive element goes up by one. But between elements 57 and 71, each successive extra electron adds to an *inner* electron shell; the outer electronic configuration remains virtually unchanged and so they rightly all belong in the same group of the periodic table. A similar situation crops up in the next period, also in Group III, with element 89, actinium. This series of elements, called the actinides, includes uranium and all the elements synthesized by man in the last thirty years or so.

Uranium

Uranium has been known since 1789. It occurs, as the rather unusual oxide U_3O_8, in the ore pitchblende. In 1896, the French physicist Henri Becquerel discovered that uranium is radioactive. There followed the various theories on atomic structure, man-made radioactivity, and atomic fission, leading to the atomic bomb and nuclear reactors. Today uranium is used as a fuel.

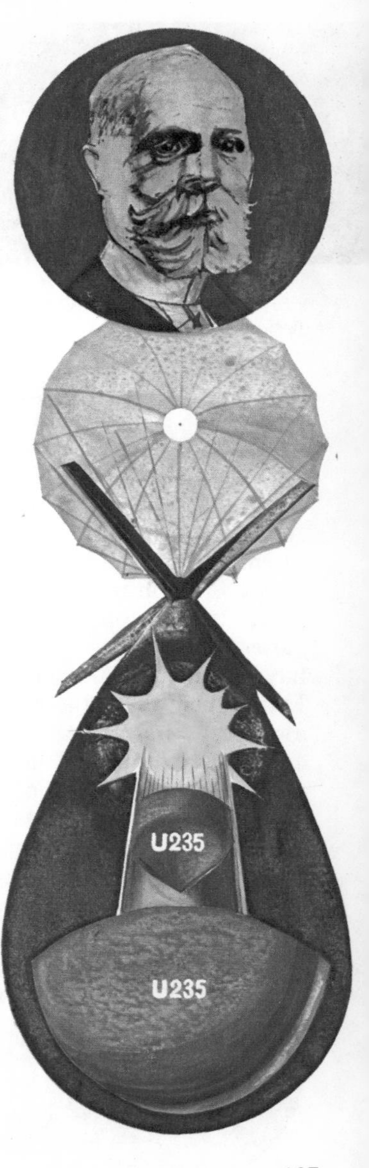

Henri Becquerel discovered the radioactivity of uranium in 1896. Forty-nine years later uranium fission was used in the first atomic bombs.

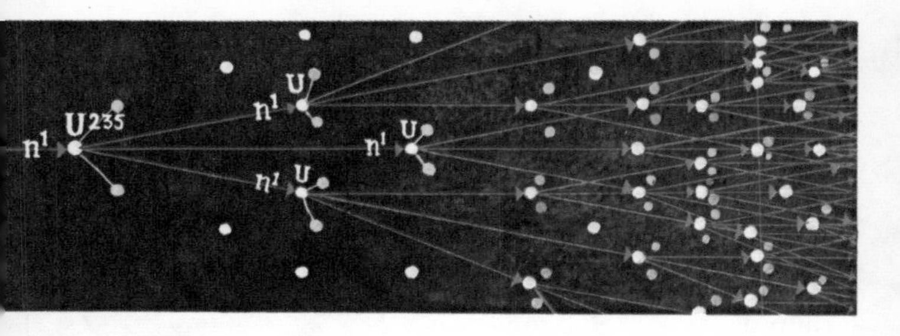

(Top) In a nuclear chain reaction, the neutrons produced by the fission of one atom go on to produce further disintegrations.

(Bottom) Nuclear fission is harnessed in a nuclear reactor, where the heat evolved is used to generate steam for powering turbines. Such a reactor can power a ship or submarine that needs refuelling only every few years.

Uranium fission

Atoms of uranium, like those of other elements, consist essentially of a collection of electrons, protons and neutrons. Each uranium atom has ninety-two electrons and ninety-two protons. The number of neutrons in the nucleus varies, giving rise to several isotopes. Atoms of uranium-235 have 143 neutrons, whereas those of uranium-238 have 146. Left to themselves, these isotopes *decay* by giving off mainly alpha rays and gamma rays (radioactivity) and eventually end up as isotopes of the element lead. U-235 decays with a half-life of 700 million years to Pb-207, and U-238 goes to Pb-206 with a half-life of 4500 million years. The long times taken for these isotopes to decay completely give ways of estimating the age of the Earth in terms of the rocks

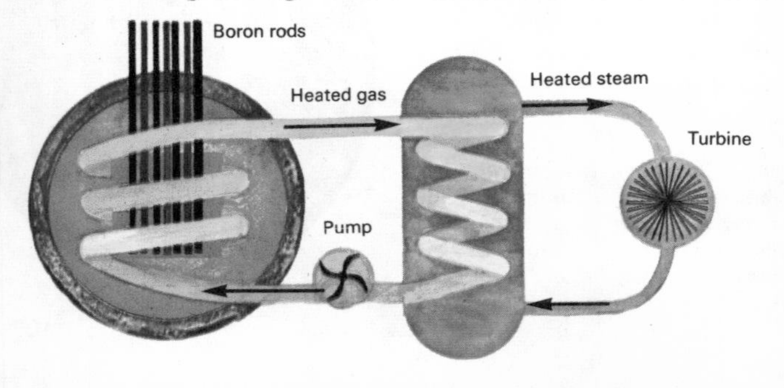

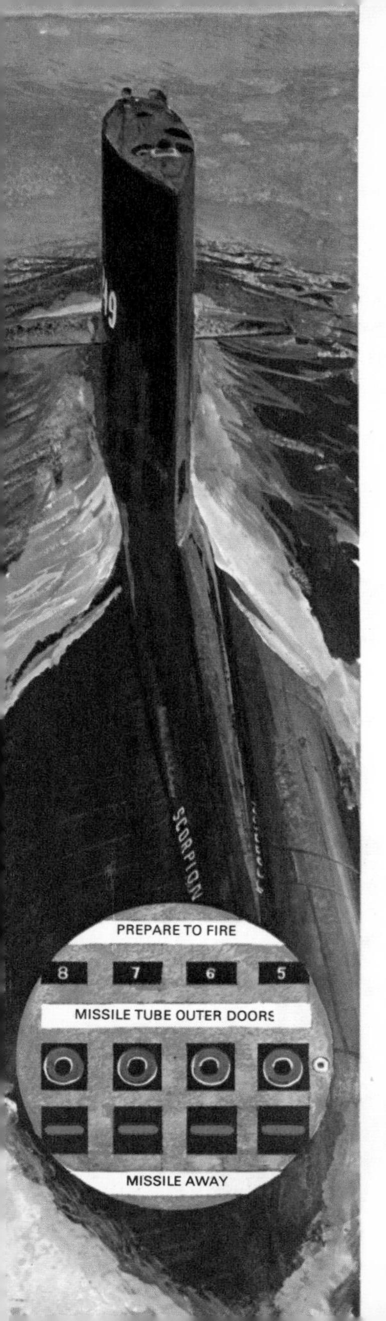

containing uranium minerals.

If a high-energy neutron is fired into the nucleus of an atom of uranium-235, the atom does not go on slowly releasing radioactive energy. Instead it splits into two roughly equal parts and releases a vast amount of energy at once. In addition to energy, it also releases three more fast (high-energy) neutrons. These can split other nearby U-235 nuclei, releasing more neutrons, and so on in a chain reaction. The splitting of atoms is called nuclear fission, and the result of an uncontrolled chain reaction is an atomic bomb.

In a nuclear reactor, energy from the fission of uranium or any other fissile element is released under control. The heat may be used to raise steam for driving a turbine which, in turn, may power a ship or an electric power generator. Controlling a chain raction is a matter of controlling the number of fast neutrons available to continue the chain. To do this, rods of a material such as boron or cadmium which 'mop up' neutrons are pushed in and out of the reactor.

THE METALLOIDS

All the elements described in the last three sections are undoubtedly metals. Physically most of them are dense and shiny, and are good conductors of heat and electricity. Chemically they form salts, generally becoming positive ions in the process, and have basic oxides. But there is a small group of elements that, while definitely not non-metals, are not sufficiently metallic to be included with the metals. These are often called the metalloids. In recent years, they have gained tremendously in importance because their in-between electrical properties make them the starting materials for semiconductors, the key substances of transistors and the whole of modern microminiature electronics.

Arsenic and antimony

Like phosphorus, the previous element in their group, arsenic and antimony exist in several allotropic forms. The yellow form is non-metallic and soluble in organic solvents, whereas the grey form is more metallic. The 'arsenic' or 'white arsenic' used by gardeners and poisoners is actually arsenic trioxide, As_4O_6. The lethal dose is a tenth of a

The Marsh test for arsenic, devised by a British chemist to detect arsenic in the remains of a person who may have died from arsenic poisoning, produces a metallic mirror of arsenic on a glass tube.

gram, but the substance is now very difficult to obtain and easily detectable. Small amounts of antimony are alloyed with lead to increase its hardness.

Both elements form gaseous hydrides called arsine, AsH_3, and stibine, SbH_3, which show strong chemical resemblances to ammonia and phosphine. They are made by the action of nascent hydrogen on any arsenic or antimony compound. At red heat, they decompose to their constituent elements. The formation and decomposition of arsine is the basis of the Marsh test for arsenic. The substance to be tested is mixed with a little zinc and dilute acid in a flask fitted with a length of horizontal glass tubing (special arsenic-free zinc must be used). The glass tube is heated with a bunsen flame. Any arsine passing the heated region forms a metallic mirror or arsenic inside the tube. The hydrogen also formed may be burnt off at the end of the tube. One, two, or even all three hydrogen atoms of arsine (and stibine) may be replaced by alkyl or aryl groups to give organic arsines (and stibines). These substances, the first organo-metallic compounds to be made, are extremely poisonous and are used in medicines, insecticides, and war gases.

Germanium

Like silicon, the element preceding it in its group, germanium is a semiconductor. The atoms of a normal metallic conduc-

Compounds of arsenic are used as drugs and as insecticides.

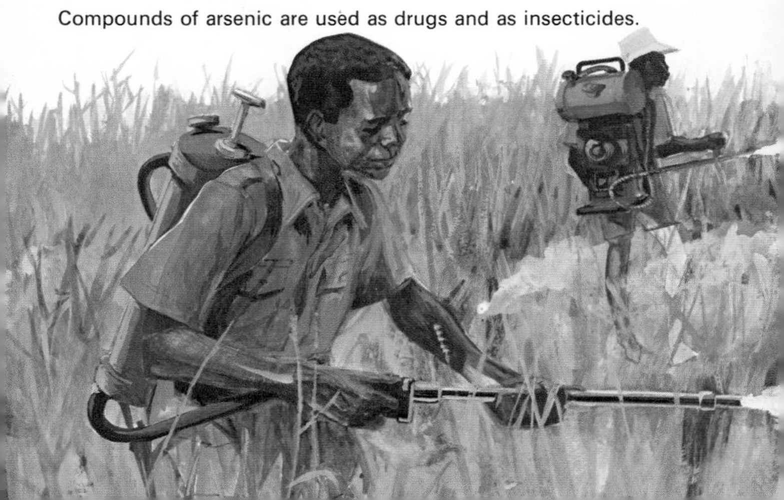

tor have plenty of 'free' electrons available for carrying electricity. The atoms of an insulator have no 'free' electrons and hence will not conduct at all. Germanium has a few 'free' electrons; at ordinary temperatures it is a poor conductor or, as it is now called, a semiconductor. But when germanium is heated, many more electrons become available to carry current and it is a good conductor.

There are other ways of making germanium a good conductor. If a very small amount of another element such as antimony or arsenic (which have one more valence electron than does germanium) is added, the number of 'free' electrons increases. Because of this excess negative charge, the resulting substance is called an *n*-type semiconductor. Alternatively, germanium may be doped with (have added to it) traces of an impurity element, such as boron or indium, which has one fewer valence electron. The effect is to create an excess of positive charges (called 'holes') for conduction and make a *p*-type semiconductor.

A piece of *n*-type or *p*-type germanium with a point

Germanium was one of the original semiconductor materials used in making transistors.

contact (like the old-fashioned cat's-whisker) makes a good diode, as does a slice of each type joined together. Transistors use a sandwich of *n*-type germanium between two pieces of *p*-type, or vice versa.

Chemically germanium is more reactive and metallic than silicon, though not so metallic as tin. It occurs in certain ores of silver and zinc. It is Mendeleeff's missing element ekasilicon, whose existence he predicted in 1869.

Selenium

Selenium is another element that apparently cannot make up its mind whether or not to be a metal. It exists as several allotropes, the non-metallic ones resembling sulphur. The grey allotrope is normally a poor conductor of electricity, but its conductivity increases when light shines on it. This photoelectricity is used in photoelectric cells for photographers' light meters and in the great 'wings' of solar cells on space craft. Hydrogen selenide, H_2Se, smells even worse and is more poisonous than hydrogen sulphide.

Selenium gives off electrons when struck by light. It is used in photoelectric cells and for satellite solar panels.

QUALITATIVE ANALYSIS

In the previous sections, the chemistry of the elements and their compounds has been described. From time to time a 'characteristic' reaction has been mentioned. For example, lead gives a bright yellow precipitate with potassium iodide; sulphates give a white precipitate with barium chloride; cadmium sulphide is yellow or orange. Armed with a knowledge of all such reactions, a chemist can identify the elements in a substance or mixture of substances. This is the function of qualitative analysis.

If he had a remarkable memory and an almost inexhaustible supply of reagents, a chemist could take the substance for analysis – the 'unknown' substance – and try to make it undergo one or more of the hundreds of known reactions. But to save time and mistakes, he would do much better to follow a systematic approach worked out over the years by analytical chemists. There are various systems and they are generally set out in the form of tables. The chemist follows through a set of instructions, performing a series of reactions in a special order. The underlying principle of a system is the use of tests which successively eliminate alternative groups of substances. If he has only a very little of the unknown substance to work on, he must use one of the methods of microanalysis. If he has plenty, which to the analytical chemist means 5 to 10 grams or about half a teaspoonful, he will probably use a method of macroanalysis.

Macroanalysis

A typical system of inorganic macroanalysis is divided into three parts: preliminary tests, tests for acid radicals, and tests for metals. It is not intended here to give a detailed set of analysis tables; instead the general principles and some specific techniques of a system will be followed.

For the first test in qualitative analysis, the chemist heats some of the dry substance in a test tube. The evolution of an alkaline gas, which turns damp litmus paper blue, indicates the presence of an ammonium salt; a blue-green substance which turns black on heating is probably a copper salt.

Preliminary tests

In the very first test, the chemist heats a little of the substance in a dry, hard-glass test tube. The substance may change colour, a gas may be evolved, or a sublimate may be formed on the cooler part of the tube. A colour change probably accompanies the decomposition of a salt to a metal oxide, and the chemist calls on his knowledge of the colours of such oxides. The oxides of copper, manganese and nickel are black; those of tin, bismuth and lead are yellow, and that of cadmium is brown. Zinc oxide is also yellow when it is hot, but turns white on cooling. Iron oxide is brown when cold, turning reddish-black when hot.

If a gas or vapour is evolved, the chemist must identify the gas by its smell or by using a chemical test. Many substances evolve water vapour, mainly because they are damp, although some crystals lose chemically attached water of crystallization. The chemist tests the steam issuing from the test-tube with a piece of damp litmus paper. If the vapour is alkaline (ammonia, which turns litmus paper blue), the substance must be an ammonium salt; if it is acid, it is probably a heavy metal salt of a mineral acid (for example,

The flame test, in which a little of the substance is burnt in the flame of a bunsen burner, can reveal the presence of various metals. A sheet of blue glass is used to mask the bright yellow flame of any sodium present.

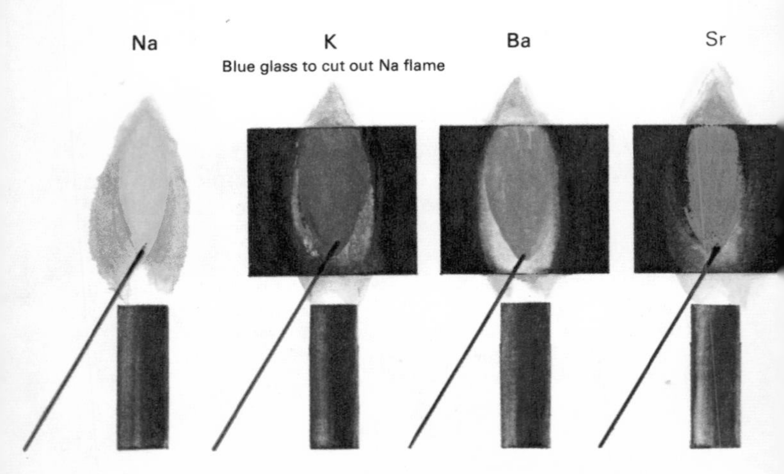

ferrous sulphate evolves sulphur dioxide and trioxide on heating). Other gases give other information, mainly about the acid radicals: oxygen indicates chlorate or nitrate, carbon dioxide indicates carbonate, iodine indicates iodide, and so on.

A white sublimate on the tube is formed by ammonium chloride, some mercury salts, and arsenic trioxide. Mercury and arsenic form grey metallic sublimates. Sulphur (yellow) and iodine (black) also sublime.

Properly carried out and interpreted, this simple test can convey much information to an experienced chemist. He would then go on to do a flame test, putting a little of the substance on a platinum wire moistened with hydrochloric acid and heating it in a luminous bunsen flame. Certain metals give characteristic flame colours.

In many reactions described in earlier chapters, carbon is used as a reducing agent. This is the point of the next test in which a little of the unknown substance is mixed with sodium carbonate and heated on a charcoal block. The alkaline earth metals (magnesium, calcium, strontium, barium) give a white alkaline residue of their oxides. Aluminium, arsenic, and zinc also give white deposits of their oxides but they are not alkaline and the zinc one is yellow when it is hot. The oxides of lead and bismuth (yellow) and of antimony (white) are completely reduced to

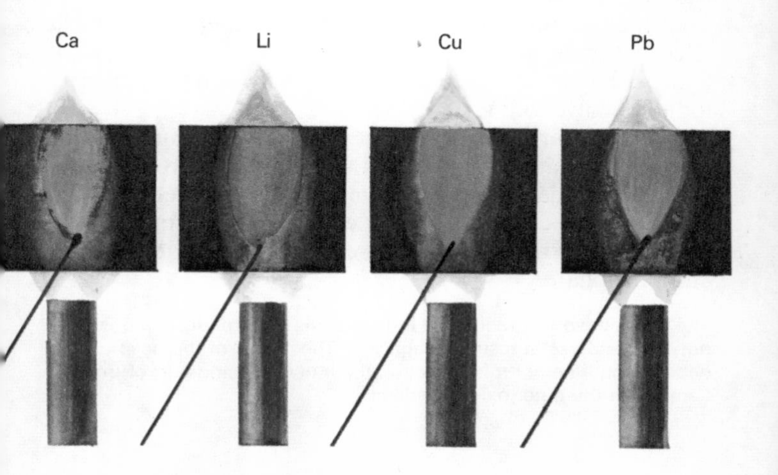

globules of metal, as are those of iron, nickel, silver and tin. Some of the white oxides formed on the charcoal block can be sorted out by adding a drop or two of cobalt nitrate solution and heating strongly. Aluminium forms a bright blue residue, magnesium a pink residue, and zinc a green one.

The final preliminary test is the borax bead test. A loop of platinum wire is made red hot, dipped into borax (sodium borate) and re-heated to give a transparent glass-like bead. A few grains of the unknown substance are added and the bead heated again. Certain metals give beads of characteristic colours, depending on whether the bead is heated in the inner reducing flame of the bunsen or in the outer oxidizing flame. The test is particularly useful for distinguishing between the metals iron, cobalt and nickel. The colours illustrated here are of the beads when they are cold.

Tests for acid radicals

The next stage in a systematic analysis is to look for acid radicals – chloride, nitrate, sulphate and so on – in the unknown substance. To do this, the chemist observes the action of various standard reagents. He first adds some dilute sulphuric acid to a small sample of the substance. Fizzing and the evolution of carbon dioxide indicates a carbonate, brown nitrous fumes a nitrite, the stench of hydrogen sulphide a sulphide, and the choking smell of sulphur dioxide a sulphite.

In the second test in this section, the chemist notes the effect of concentrated sulphuric acid. Chlorides evolve hydrogen chloride, nitrates give off brown nitrous fumes, iodides evolve iodine, and so on.

The addition of dilute nitric acid and silver nitrate solution gives precipitates with halides (the chloride precipitate is white, bromide is cream coloured, iodide is yellow). The same reagent can also be made to reveal the presence of more obscure radicals, such as chlorate, bromate, arsenate and phosphate.

The borax bead test requires a little experience to perform properly, but can yield useful results in analysis. The colour of the bead depends on whether it is heated in the inner reducing part of the flame or in the outer oxidizing flame.

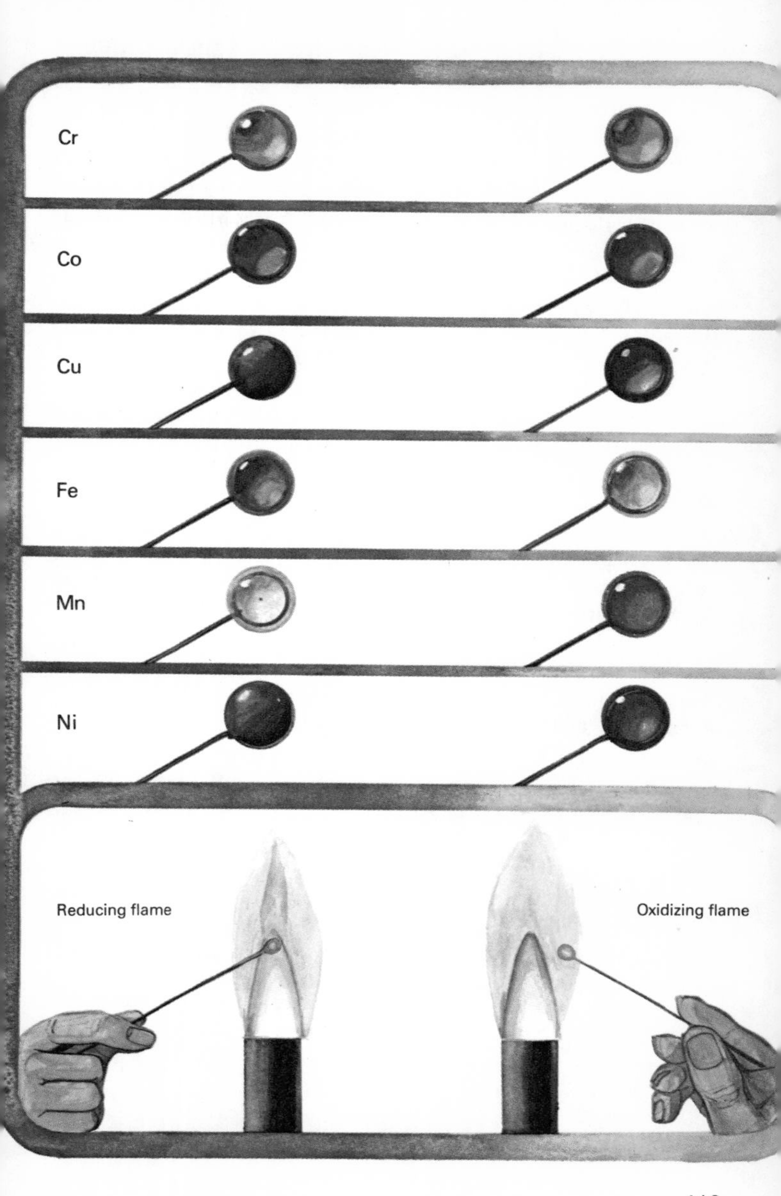
Cr
Co
Cu
Fe
Mn
Ni
Reducing flame
Oxidizing flame

The formation of a white precipitate after the addition of hydrochloric acid and barium chloride solution to the unknown substance indicates the presence of a sulphate.

Nitrates are revealed by the brown ring test, which may be carried out in various ways. A good method is to dissolve some of the unknown substance in a little water and put the solution in an evaporating basin. A crystal of ferrous sulphate is placed in the centre of the liquid and a few drops of concentrated sulphuric acid cautiously dropped on to it. The formation of a brown ring indicates the presence of a nitrate. An alternative test for nitrate relies on its reduction to ammonia. A little sodium hydroxide solution and Devarda's alloy (aluminium, copper and zinc) is added to the unknown substance. The evolution of ammonia, detected by its smell and action on litmus, indicates a nitrate or nitrite.

This latter test fails if the unknown substance contains an ammonium salt, so this should be tested for by adding sodium hydroxide solution and warming. Under these conditions ammonium salts yield ammonia gas. If an ammonium salt is present, it should be completely decomposed with alkali before testing for nitrate.

Tests for metals

Metals generally exist in the unknown substances as metal ions. Most systems of analysis depend on precipitating them as carbonates, hydroxides or sulphides from either acid or alkaline solution. The metals are then identified from the colour of the precipitate and by separate confirmatory tests. By using a sort of cascade principle, the metals are separated into groups. Any precipitate formed after the first separation reaction includes metals of analysis group I. The precipitate is filtered off, and the filtrate carried over to the second separation reaction. Any precipitate, also filtered off, includes metals of group II. The filtrate is then treated to separate group III, group IV, and so on.

Each group precipitate is then subjected to a series of tests to separate it into its component metals. Using this system, an analysis scheme can be made as comprehensive as the chemist likes. Here, in very brief outline, is a simple scheme which includes all the commoner metals. Various substances,

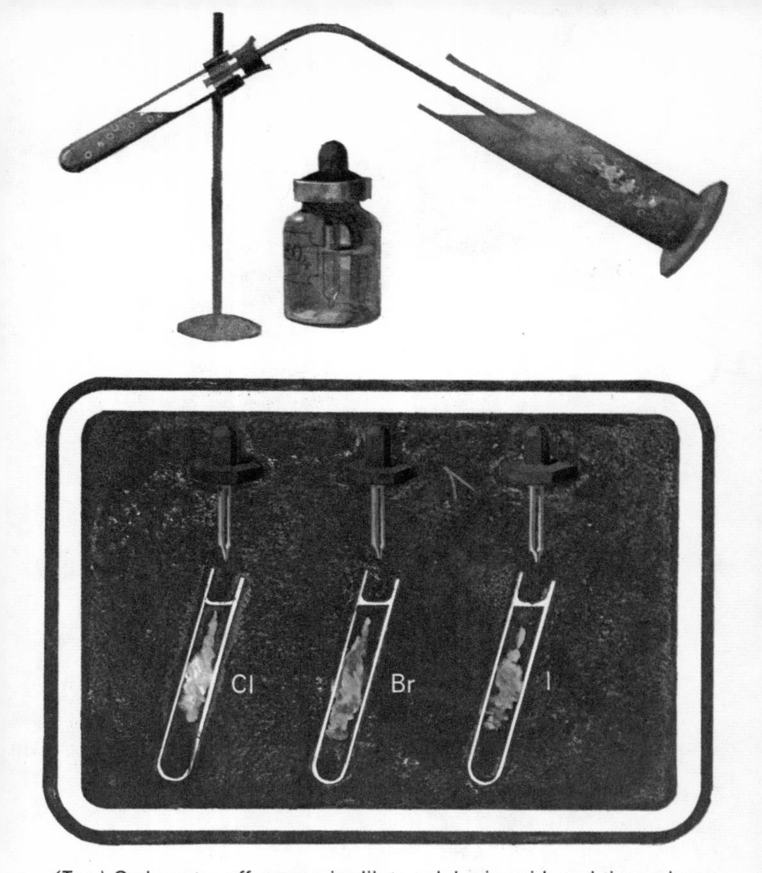

(Top) Carbonates effervesce in dilute sulphuric acid, and the carbon dioxide produced turns lime-water milky.
(Bottom) Halides give precipitates with silver nitrate solution and dilute nitric acid; chlorides give a white precipitate, bromides a creamy-white precipitate, and iodides a yellow precipitate.

such as phosphate, interfere with the scheme and must be tested for and removed if necessary.

Add dilute hydrochloric acid. A white precipitate, the chloride of lead, mercury or silver, contains the metals of analysis group I. Any precipitate is filtered off and hydrogen sulphide gas bubbled through the filtrate. A coloured precipitate, the sulphide of antimony, arsenic, bismuth, cadmium,

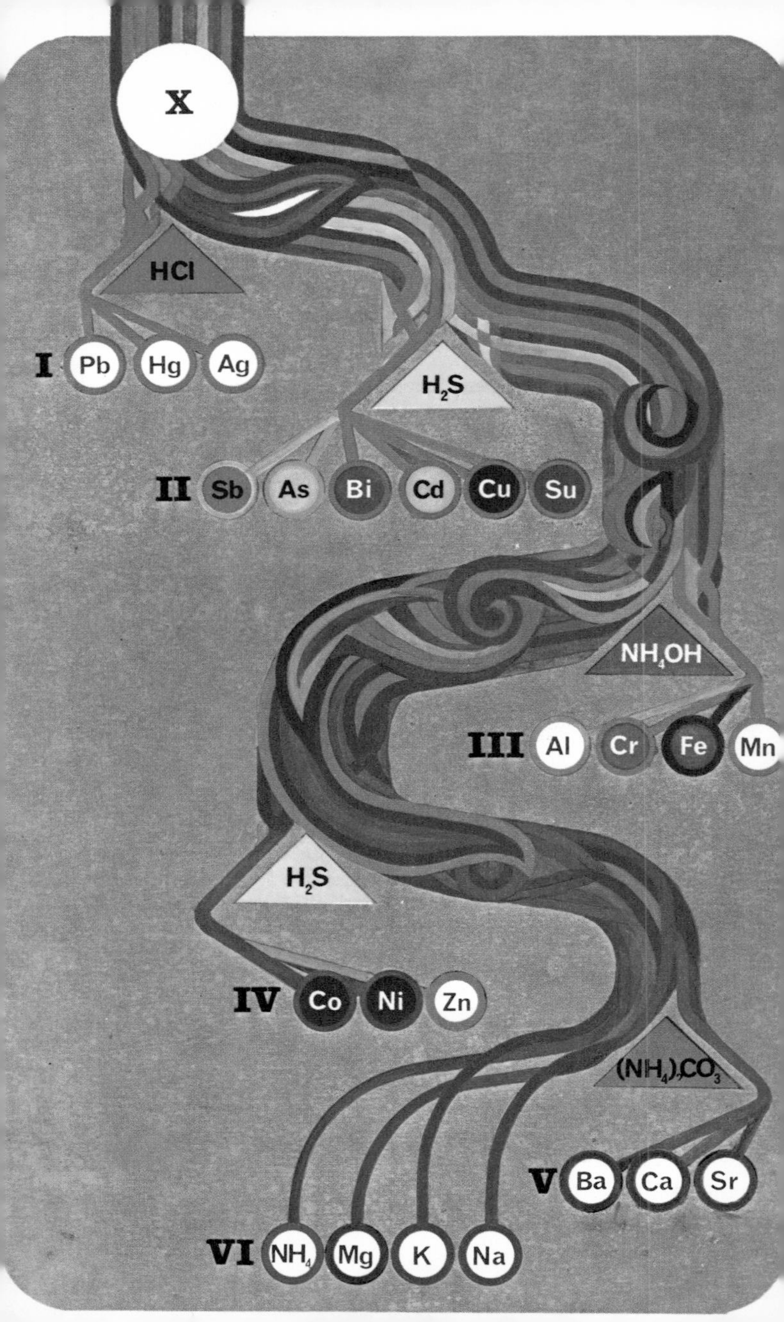

X
HCl
I
Pb
Hg
Ag
H2S
II
Sb
As
Bi
Cd
Cu
Su
NH4OH
III
Al
Cr
Fe
Mn
H2S
IV
Co
Ni
Zn
(NH4)2CO3
V
Ba
Ca
Sr
VI
NH4
Mg
K
Na

copper or tin, contains the metals of group II. These are filtered off, the filtrate concentrated by boiling and made alkaline with ammonium chloride and ammonium hydroxide. A precipitate, the hydroxide of aluminium, chromium, iron or manganese, contains metals of group III. These are filtered off and hydrogen sulphide bubbled through the filtrate. A precipitate, the sulphide of cobalt, nickel or zinc, contains the metals of group IV. Again after filtering, the solution is concentrated, made alkaline, and ammonium carbonate added. A precipitate, the carbonate of barium, calcium or strontium, contains the metals of group V. Finally if there is anything left, it must be in group VI which includes magnesium, potassium and sodium. The ammonium ion is also placed in this group.

The diagram on the opposite page shows how the 'cascade' idea works; the spill over from the first separation is the starting material for the second, and so on. The colours shown are typical, although in practice they may vary depending on the fineness of the precipitate. Also one strongly coloured precipitate (say a black one in group II) may mask the colours of any other precipitates formed at the same time. For this reason, each group precipitate has its own separation scheme.

Microanalysis

The sort of qualitative analysis scheme just described will detect about two dozen metals and nearly as many acid radicals, providing the analytical chemist has several grams of starting material. If he has less than a gram, he must use one of the systems of semimicro- or micro-analysis. These have similar separation procedures using the 'cascade' principle on a very much reduced scale. Special miniature apparatus is used; small test tubes holding only 3 or 4 ml and tiny beakers that hold about twice that volume. Solutions are not poured but manipulated using eye-droppers; each reagent bottle has its own dropper. For speed and for handling the minute amounts of precipitate, a centrifuge

The qualitative analysis scheme for metal radicals is like a cascade, with various reagents precipitating groups of metals at each stage.

is used to fling the precipitate to the bottom of its tube. The clear liquid above (corresponding to the filtrate in a normal filtering operation) may be drawn off in a dropper and transferred to another tube to continue the analysis. Miniature crucibles, burners and water baths complete the semimicro kit. Many of the characteristic colour reactions are carried out on glass slides or on a porcelain plate using single drops of chemicals. Microanalysis tables generally include many of the less common elements. The new techniques must be learned and practised, and an experienced chemist can analyse a mixture containing six metals in an hour or two.

Spectroscopy

Using a spectroscope, a chemist can analyse a mixture of metals in a few seconds. We have already seen in the flame test how certain metals burn with a flame of characteristic colour. All elements give off light when they incandesce and burn, though not always in the visible part of the spectrum. Also the white light consisting of mixtures of colours given off by an incandescent metal always looks much the same to

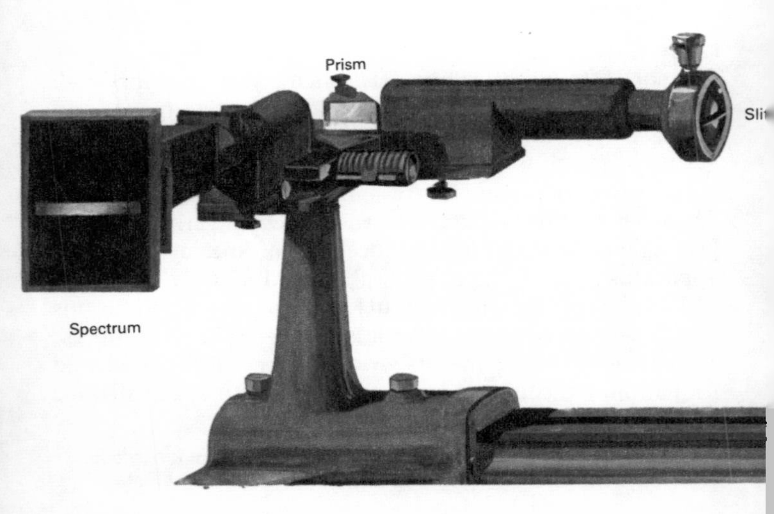

our eyes. So the chemist uses a spectroscope to split the white light into its component colours or spectrum. The spectrum of each element is characteristic – a sort of fingerprint that the chemist can use to identify it. A small portable version of the instrument, called the direct vision spectroscope, can be used in the course of ordinary inorganic analysis. For example, it is useful for sorting out calcium, strontium and lithium, which all give a red flame in the flame test.

If a metal is made the target in an X-ray tube and the tube operated at very high voltages, the X-rays produced show peaks at certain frequencies. These frequencies are characteristic of the metal (another fingerprint for the analyst) and are the basis of X-ray spectroscopy.

Organic analysis

So far in this section only methods of inorganic qualitative analysis have been considered. But the substance given to

Each metal has a characteristic spectrum: tiny quantities can be detected and identified using a spectroscope.

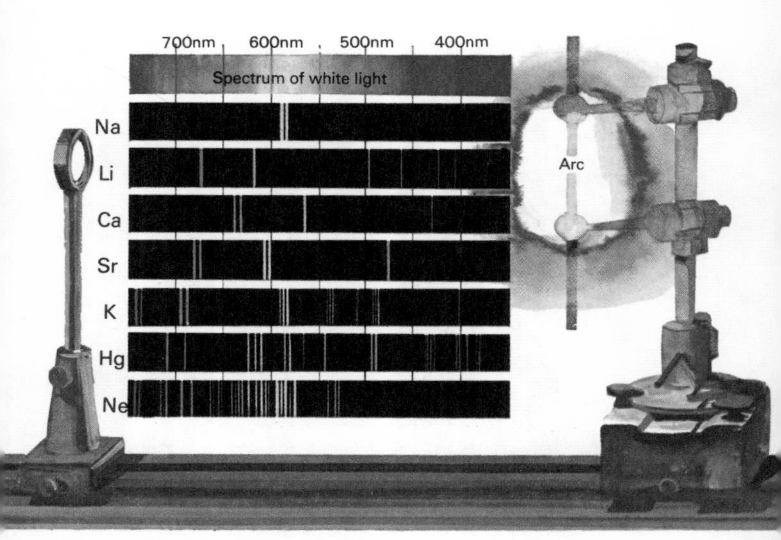

a chemist to analyse may be organic in nature. So chemists have also devised a systematic way of analysing organic substances. Again it is beyond the scope of this book to give precise experimental details, but the following brief account illustrates the general principles involved.

First of all the analyst decides whether he is dealing with a single substance or a mixture of substances. If he suspects he has a mixture, he tries to separate it into its components by physical means such as solvent extraction (when one component of the mixture dissolves in a solvent whereas the remainder does not), distillation (for separating two liquids of different boiling points), and so on. If none of these methods works, the mixture must be separated by chemical means.

The chemist then goes on to look for 'non-organic' elements such as a metal, nitrogen, sulphur, or the halogens. A little of the substance is strongly heated on a crucible lid until all the carbon has burnt away. The presence of an ash

In organic analysis, 'non-organic' elements are detected in Lassaigne's sodium fusion test *(right)*. The solution from the fusion is tested with sodium nitroprusside, which gives a purple colour with sulphur *(far right)*; with ferrous sulphate, which gives prussian blue with nitrogen *(far left)*; and with silver nitrate, which gives a white precipitate with chlorine *(left)*.

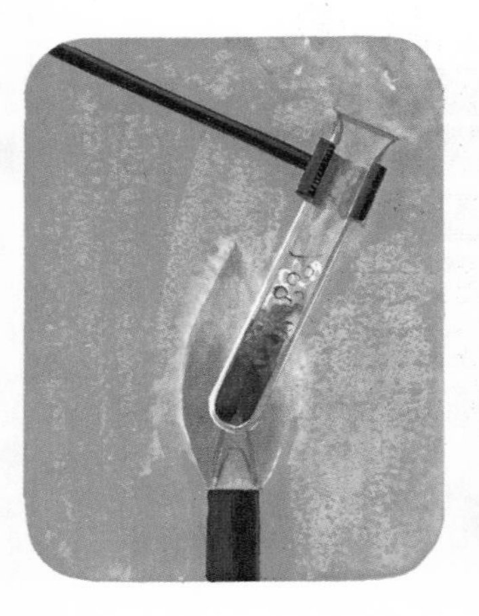

or residue indicates that a metal is present. It is identified by the methods of inorganic analysis just described.

A single test, the Lassaigne sodium fusion, detects the presence of nitrogen, sulphur, or the halogens. A little of the substance is heated to red heat with a small pellet of sodium in a test tube. The red-hot tube is plunged into some distilled water in an evaporating basin. Any nitrogen will have been converted by the fusion to sodium cyanide, any sulphur to sodium sulphide, and any chlorine (or other halogen) to sodium chloride (or other halide). So the chemist filters the solution and tests it for each of these sodium compounds.

The analyst then goes on to test for characteristic organic groupings. If the preliminary tests have revealed nothing, he assumes that the compound contains only carbon, hydrogen, and oxygen. He tests its solubility in water and ether, tests it for acidity or basicity, and tests for carbonyl groups, ester linkages, and hydroxyl groups. At the end of these tests, he should know whether or not he has an acid, an ester, an alcohol, a phenol, a base, an aldehyde or ketone, or a hydrocarbon. To find out just *which* acid or ester or whatever, he prepares a solid derivative, measures its melting point, and consults lists of tables for a positive identification.

If the sodium fusion test indicated the presence of nitro-

gen, the chemist does another series of tests to determine whether the compound is an amine, amino-acid, amide, nitrile, nitro compound, and so on. A sulphur compound is likely to be a sulphonic acid, sulphonamide, or thiol. Halogens turn up in alkyl halides and halogen substituted aromatic compounds.

Again to pinpoint the exact compound, the chemist must prepare a derivative and measure its melting point. The choice of derivative depends on the starting material and includes nitro compounds, picrates, semicarbazides, sul-

(Left) Some mixtures of organic compounds can be separated by shaking an aqueous suspension with ether, when one compound dissolves in the upper ether layer and the other remains in the water layer.

(Centre) A melting point determination requires a sample of the solid substance, and saturated solutions can be induced to crystallize by scratching the inside of the test tube with a glass rod.

(Right) To find the melting point of an organic solid, a small quantity of it is put in a capillary tube, attached to the base of a thermometer, and carefully heated in an oil bath until the solid melts.

phonates and many others. The analyst also measures the melting point or boiling point of a purified sample of the unknown substance.

Of course, organic compounds can be almost infinitely complex and some natural plant and animal products defied complete analysis for many years. Even a knowledge of all the active groupings, molecular weight and general structure may not be sufficient to pin-point a compound exactly. The existence of ordinary, geometric and optical isomers also increases the difficulty of a complete structural analysis.

But the organic chemist also has a number of physical methods he can apply. He too can use spectroscopy, measuring the *absorption* spectrum of a solution of the substance to ultra-violet or infra-red light. He can measure the refractive index of liquids and the optical activity of certain compounds. Recently X-ray methods have also come to his aid. As a result, the organic chemist can generally determine the structure of a useful substance, and armed with this knowledge devise a way of synthesizing the substance for the general benefit of mankind.

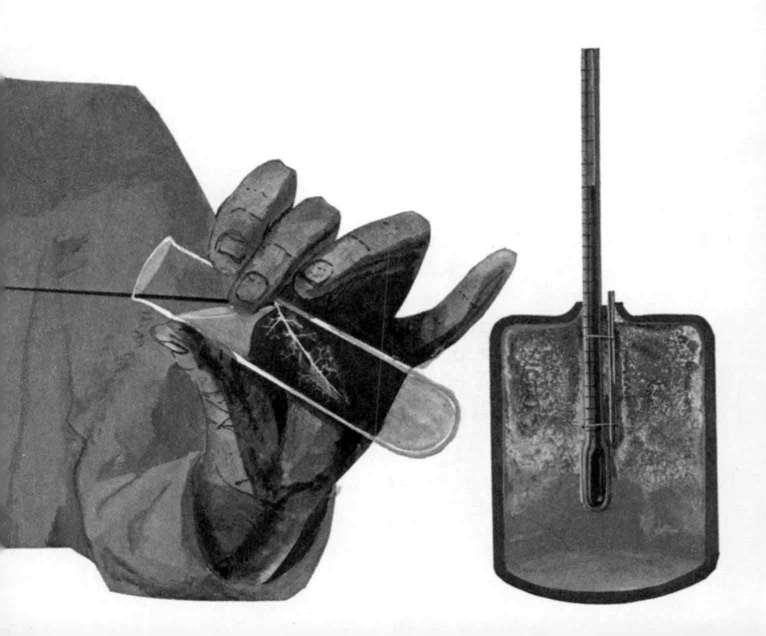

QUANTITATIVE ANALYSIS

The techniques of qualitative analysis described in the previous section tell the chemist what an unknown substance is; he uses quantitative analysis to find out the amount. He may have a dilute solution of what he knows is sulphuric acid, but he has to analyse it to find out exactly how dilute the acid is.

There are two main conventional methods of quantitative analysis. In one method, the substance to be analysed and all the reagents are made into solutions. All measurements are made in terms of volumes of liquids; the method is called *volumetric* analysis. In the second method, the element to be determined is precipitated out of a solution as an insoluble compound. All measurements are made in terms of the weights of solids; this is *gravimetric* analysis.

In addition there are other methods of analysis that make use of different properties of chemicals. These more recent techniques are particularly useful for analysing mixtures of substances. The different rates at which chemicals are mopped up by absorbent paper gave rise to the method known as *chromatography*. The differences in the ease with which various ions will change partners on an ion exchange resin (as in water softening) are used in another method. The different rates at which ions move in solution under the influence of an electric field and even the different weights of the ions themselves have all been used as way of analysing mixtures quantitatively.

Volumetric analysis

In this method, all the substances are in solution. The chemist selects a chemical reaction involving the substance to be analysed. He takes a known, precisely measured volume of one solution and measures how much of the second solution is needed to react with it. Then if he knows the concentration of either of the solutions, he can easily calculate the strength of the other one.

Volumetric analysis determines the exact quantitative composition of a substance in solution.

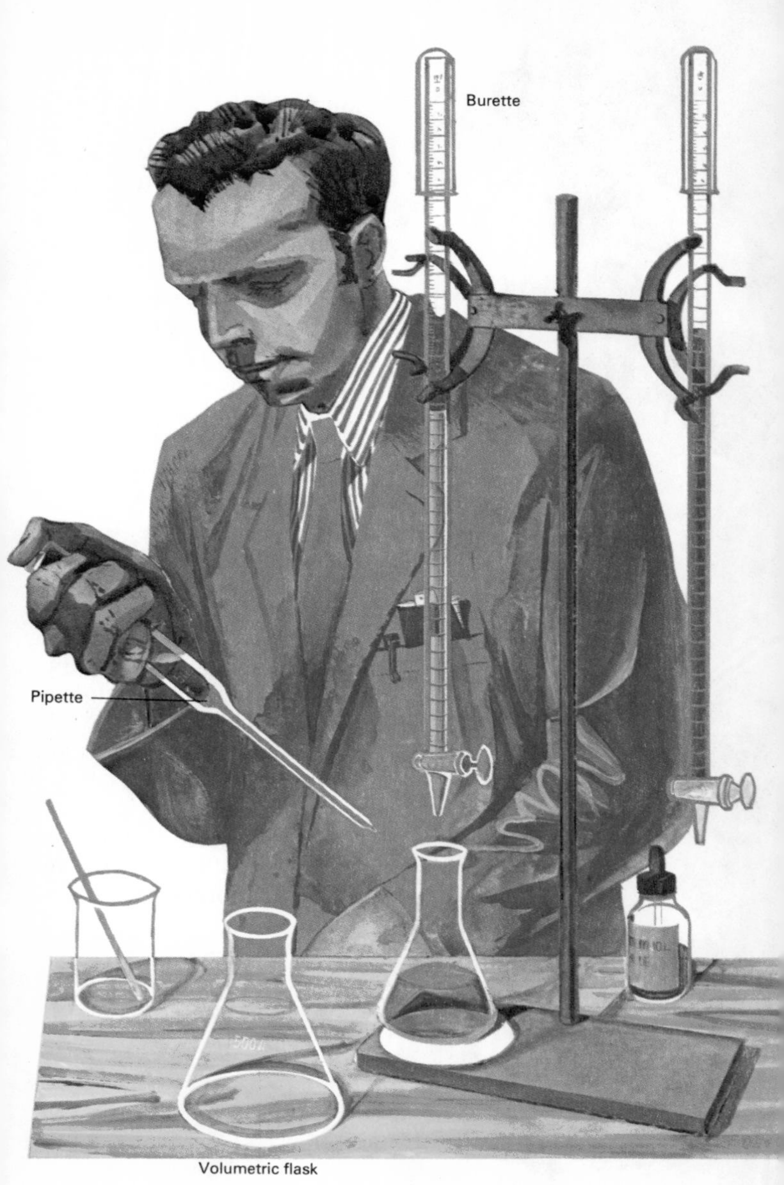
Burette
Pipette
Volumetric flask

A simple example will make the method clearer. Suppose the chemist wishes to find the strength of some dilute hydrochloric acid. He has a supply of an alkali, say sodium hydroxide solution, of known strength, say 10 per cent. He can 'watch' the reaction between an acid and an alkali using an indicator such as litmus, which is red in acid solution and blue in alkali. So he takes a known volume of sodium hydroxide solution (say 25 ml) and adds a drop or two of litmus, turning it blue. He measures this volume by sucking the solution into a precisely calibrated bulb known as a pipette. He then adds the acid solution slowly from a long graduated tube called a burette. He stops adding acid the instant the solution goes red.

The procedure just described is called a *titration*, and the instant that the reaction is complete is called

(Top) Accurate volumes of solution are dispensed using a pipette.
(Bottom) Stock solutions of known concentration are made up in a volumetric flask or, less accurately, in a graduated measuring cylinder.

the end-point of the titration. In practice, chemists do not generally measure the strengths of solutions in percentages. Instead they use normality; a normal solution contains one gram equivalent weight of a substance dissolved in one litre of solution. A normal solution is written as $1N$, a tenth-normal solution as $0{\cdot}1N$ or $N/10$, and so on.

In general in a titration, if a volume V_1 of a solution of strength S_1 neutralizes exactly a volume of V_2 of a solution of strength S_2,

$$V_1 \times S_1 = V_2 \times S_2.$$

If the chemist knows both volumes and one strength, he can easily calculate the unknown strength. In the acid-alkali titration of the example, the 10 per cent sodium hydroxide solution has a strength of $2{\cdot}5N$, expressed in terms of normality. The equivalent weight of sodium hydroxide is 40. A 10 per cent solution contains 100 gm/litre and therefore $\frac{100}{40}$ ($= 2{\cdot}5$) gram equivalents per litre.

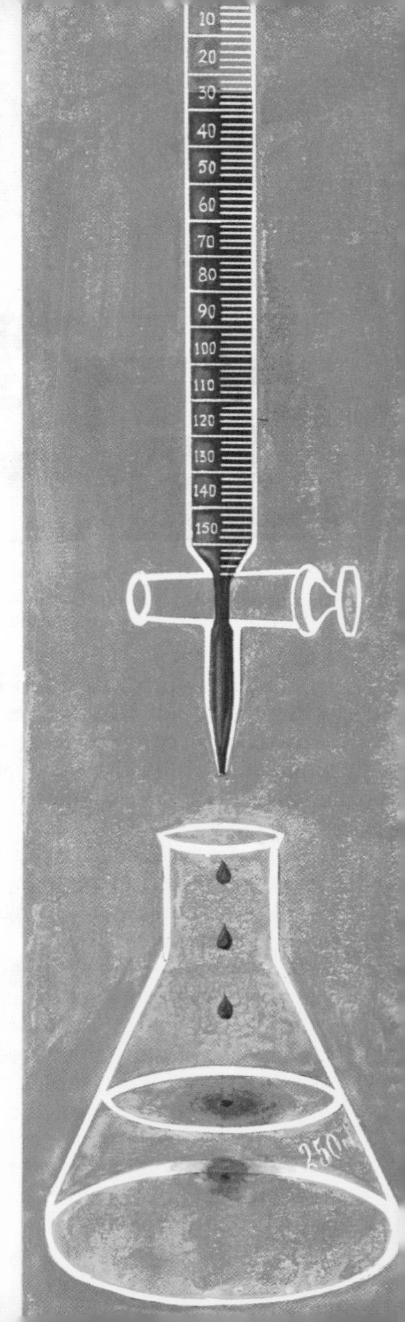

In a titration, the 'end point' of a reaction is generally detected by the colour change of an added indicator.

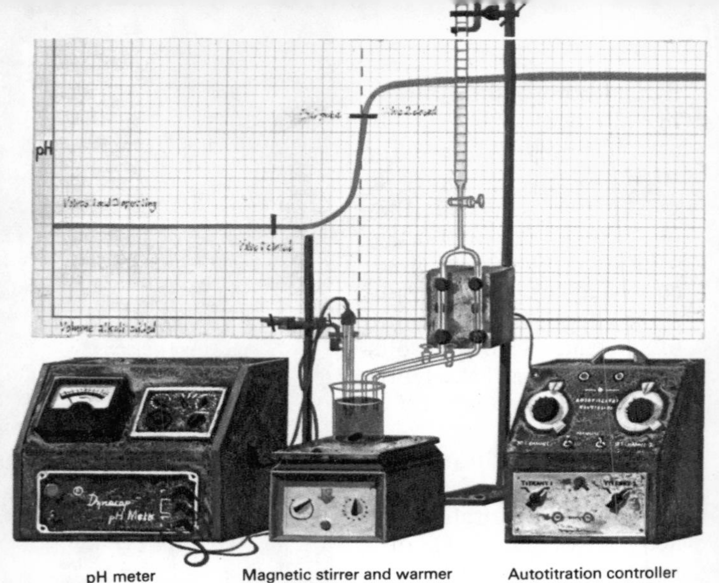

In a potentiometric titration, the 'end point' is detected by a sudden change in electrical conductivity of the solution.

The acid-alkali titration just described is not the only sort of reaction that can be used in volumetric analysis. Any reaction can be used so long as it has a detectable end-point. Generally these include all reactions with a colour change or for which there is an indicator which changes colour. For example, an oxidizing agent can be titrated against a reducing agent (the redox reaction). If the oxidizing agent is potassium permanganate, it is its own indicator because it changes colour from bright purple to colourless at the end-point.

There are other ways of detecting the end-point that monitor the changes in ionic concentration during a titration. A conductometric titration makes use of conductivity changes, a potentiometric titration follows in electrode potentials, a polarographic titration relies on the current-voltage characteristic of the solution, and so on. Most of these methods require complex apparatus and graph-plotting, and so are generally employed only when ordinary methods cannot be used.

Gravimetric analysis

In the various methods of gravimetric analysis, the quantity of substance is determined by precipitating it from solution as an insoluble compound and weighing the precipitate. For example, a chloride can be determined by precipitating it as silver chloride, a sulphate by precipitating barium sulphate, and so on. Most metals can be precipitated as their hydroxides or carbonates, which may be roasted to form the oxides before weighing. Metals with soluble carbonates and hydroxides, such as the alkali metals, are precipitated as insoluble complex compounds. The success of the methods depends mainly on technique – making sure all the substance is precipitated, making sure all the precipitate is collected, making sure it is completely dry before weighing.

Gravimetric analysis relies on weighing for its accuracy. In this sequence, ammonium hydroxide is used to precipitate ferric hydroxide, the solid is filtered off through a Buchner funnel, the filtrate and filter paper are strongly heated to burn away the paper and convert the hydroxide to oxide, and finally the solid iron oxide is weighed.

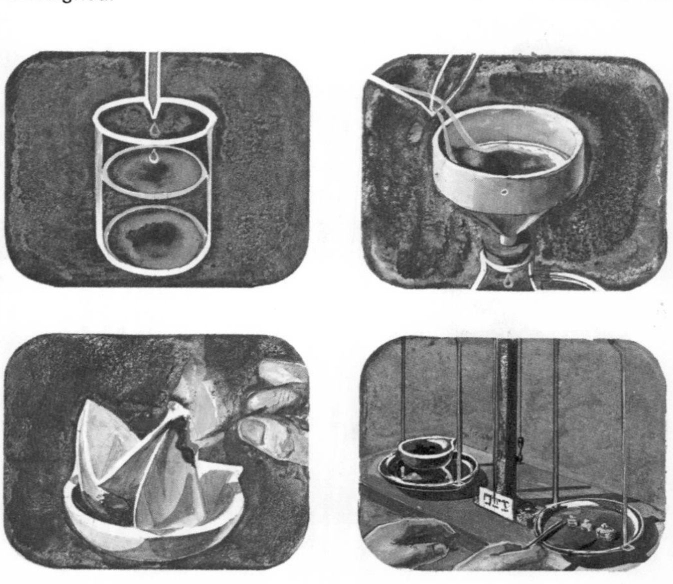

RUBBER AND PLASTICS

So far in this book theoretical, inorganic, organic and analytical chemistry have been briefly considered. For the remainder some of the major chemical industries will be dealt with (heavy chemicals such as acids and alkalis are discussed under their inorganic chemistry). The first industries to be looked at developed from a natural product, rubber, and then went on to use chemical techniques to synthesize a whole range of man-made copies of natural products, the plastics. All these products have one thing in common; they are made up of huge molecules.

Natural rubber occurs as a sticky white fluid called latex in the tissues of certain trees. The milky latex is collected by tapping the tree and is treated with ammonia and formic acid to form a spongy mass called *coagulum*. This is rolled into sheets and smoked over a fire.

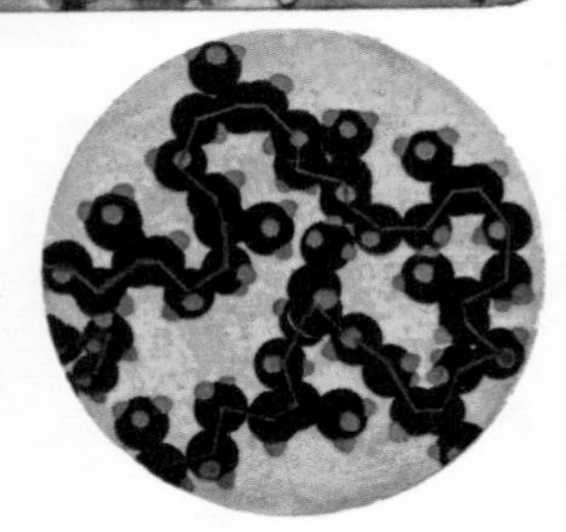

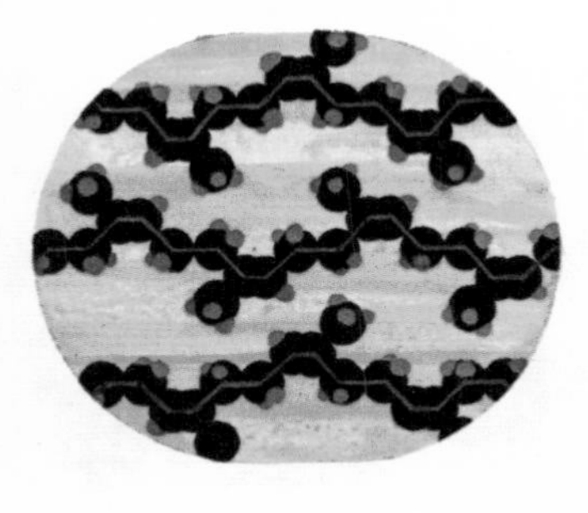

Natural rubber is tapped as latex from beneath the bark of a rubber tree *(top)*. Molecules of natural rubber *(centre)* are kinked; they straighten when the rubber is stretched *(bottom)* and spring back when released.

To toughen rubber and make it keep its elasticity over a wide temperature range, it is vulcanized. In a process largely unchanged since it was discovered by Charles Goodyear in 1839, the rubber is mixed with about 3 per cent of sulphur and an accelerator (catalyst) and heated to about 150°C. Rubber for tyres also has carbon added.

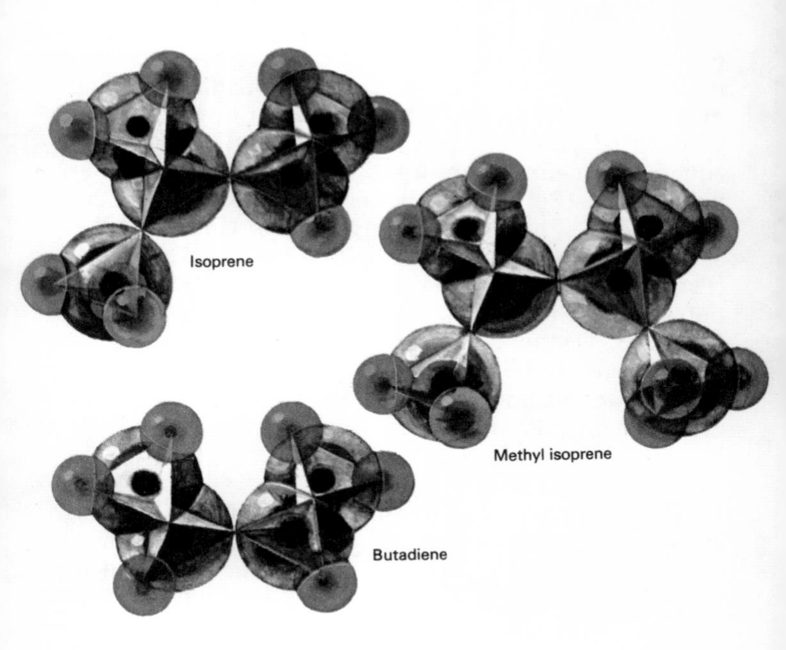

When substances used for making synthetic rubber are polymerized, their molecules form long, kinked chains like those of natural rubber.

Chemically rubber is a polymer of the olefine hydrocarbon isoprene. $CH_2{:}CH{\cdot}C(CH_3){:}CH_2$. Thousands of isoprene molecules are joined end to end to form the chain-like molecules of rubber. But the chains are not straight; they are jumbled up like a handful of tangled lengths of string.

In the earliest attempts to make synthetic rubber, chemists tried to make polymers of isoprene. Success did not come until they used the closely related compounds methyl isoprene and butadiene. Not until 1955 was rubber synthesized from isoprene.

Polymers

Synthetic rubber and other plastics are polymers. Their molecules consist of thousands of simple units joined together. Their chemical names reflect this fact, although manufacturers often shorten the names or invent new ones to describe plastics. Some common plastics, with their American or European trade names (in brackets), are polyethylene (Polythene, Alkathene), polytetrafluoroethylene (PTFE, Fluon, Teflon), polyvinyl chloride (PVC, Vinyl), polystyrene (Styroflex), polymethyl methacrylate (Perspex, Plexiglass), polychlorobutadiene (Neoprene), polyvinyl alcohol (PVA), polyethylene glycol terephthalate (Terylene, Dacron), polyamides (Nylon). There are also alkyd resins, Bakelite and Melamine. They are made by one of two methods: addition polymerization and condensation polymerization.

Addition polymers

Many plastics, including the first seven on the above list, are made from double-bonded monomers by an addition reaction. Consider the polymerization of ethylene, $CH_2:CH_2$. In the first stage, called the initiation phase, the

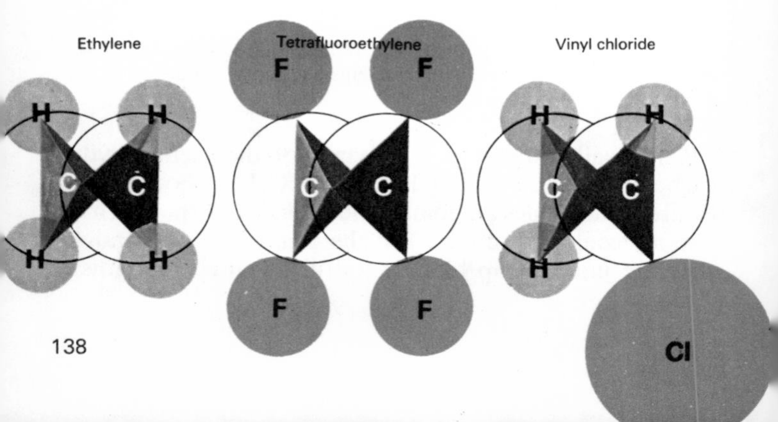

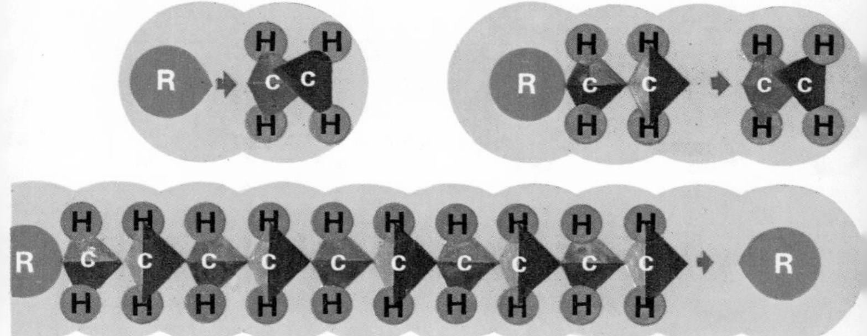

double bond is broken by heat, light, X-rays or, more commonly, by a source of free radicals. Organic peroxides, of general formula R·OO·R, are used to generate free radicals R·. These initiate the chain reaction by reacting with a molecule of the monomer to form a bigger free radical $R{\cdot}CH_2CH_2{\cdot}$. Then in the growth phase, this free radical attaches itself successively to more and more ethylene molecules to form a long chain $R(CH_2)_nCH_2{\cdot}$. Finally in the termination phase, this long chain free radical reacts with another free radical to put an end to the chain and stop the reaction, forming $R(CH_2)_nR$.

(Top) In addition polymerization, thousands of monomer molecules join end to end to form long-chain molecules of a plastic.

(Bottom) Many polymers are based on ethylene or its related compounds, all of which retain a carbon-carbon double bond.

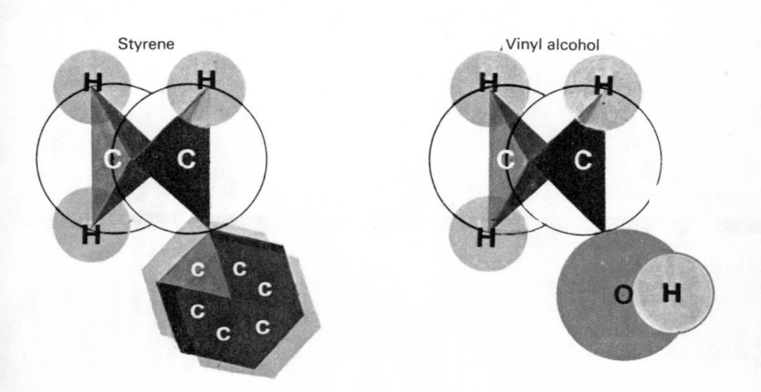

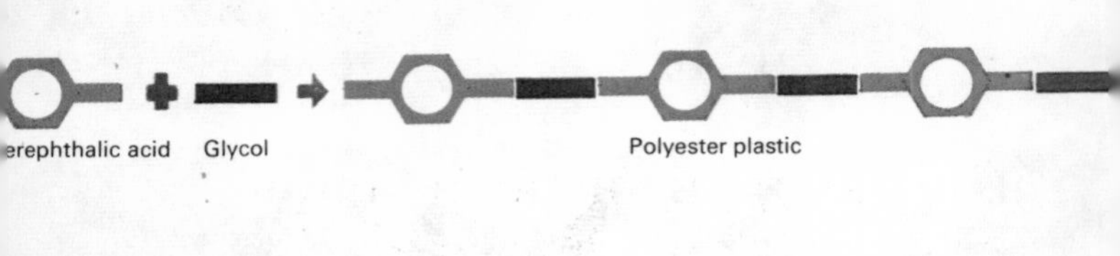

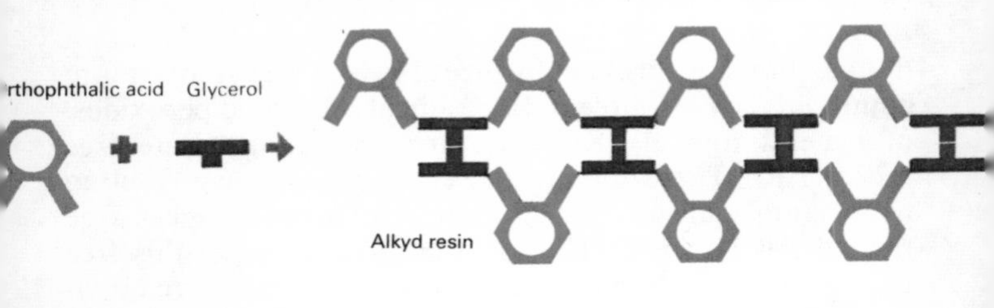

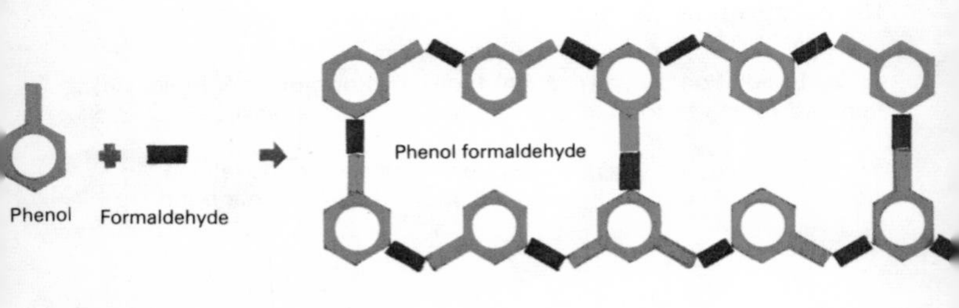

Adipic acid Hexamethylene diamine Nylon

Terylene Dacron

Paints

Bakelite

Nylon gears

Condensation polymers

Another group of polymers is made by condensation reactions in which two molecules (generally with no double bonds) come together and react. They join together with the elimination of a simple compound such as water. For instance an alcohol such as ethylene glycol will condense with an acid such as terephthalic acid to form a polyester resin. Glycerol condenses with orthophthalic acid to form an alkyd resin.

The earliest plastics were made by condensing formaldehyde with phenol. The result, a phenolic resin, is better known as Bakelite after its discoverer Leo Baekeland who first carried out the reaction in 1909. Similar condensation polymers are urea-formaldehyde resins and amines will condense with acids to give polyamides such as Nylon. The original form of Nylon was made by condensing adipic acid with hexamethylene diamine.

Condensation polymers, formed when two simple reactants condense in their thousands to give huge chain or net-like molecules *(left)*, include Terylene, alkyd resins for paints, Bakelite and Nylon.

COAL AND PETROLEUM

Coal and petroleum are fossil fuels derived from the breakdown of vegetation buried beneath the ground millions of years ago. They are therefore both organic in nature, coal containing mainly aromatic compounds and petroleum containing mainly aliphatics. Both are separated into their principal constituents by distillation. These constituents then become the chief raw materials of the organic chemicals industry.

Distillation of coal yields coal gas, ammonia, coal tar and coke. Coal gas consists

Coal, an essential raw material for modern chemical industry, is today mined increasingly by automated machinery. Some products derived from coal are shown on page 144.

Another important raw material, petroleum, has the advantage that it can be pumped long distances through pipelines from the oil well to the nearest port, for shipping to an oil refinery (page 145).

mainly of hydrogen, methane and carbon monoxide, all of which are inflammable. Coal tar can be separated, again by distillation, into many useful chemicals including benzene, toluene, phenol, naphthalene and pitch. Coke is used as a fuel and to make fuel gases such as producer gas and water gas. Ammonia is converted into ammonium sulphate for use as an artificial fertilizer. The sulphuric acid for doing this may also be derived from coal.

Distillation of petroleum separates the crude oil into 'fractions' of hydrocarbons – butane (gas), octane (petrol), dodecane (kerosene), fuel oil, lubricating oil, paraffin wax and tar. The high molecular weight compounds in oil are then broken down into the more useful paraffins of petrol and kerosene. This is done with the aid of a catalyst (catalytic cracking) or by the action of heat (thermal cracking). The low molecular weight paraffins are starting materials for making solvents and plastics. Wax may be removed from crude oil by extraction in solvents such as nitrobenzene.

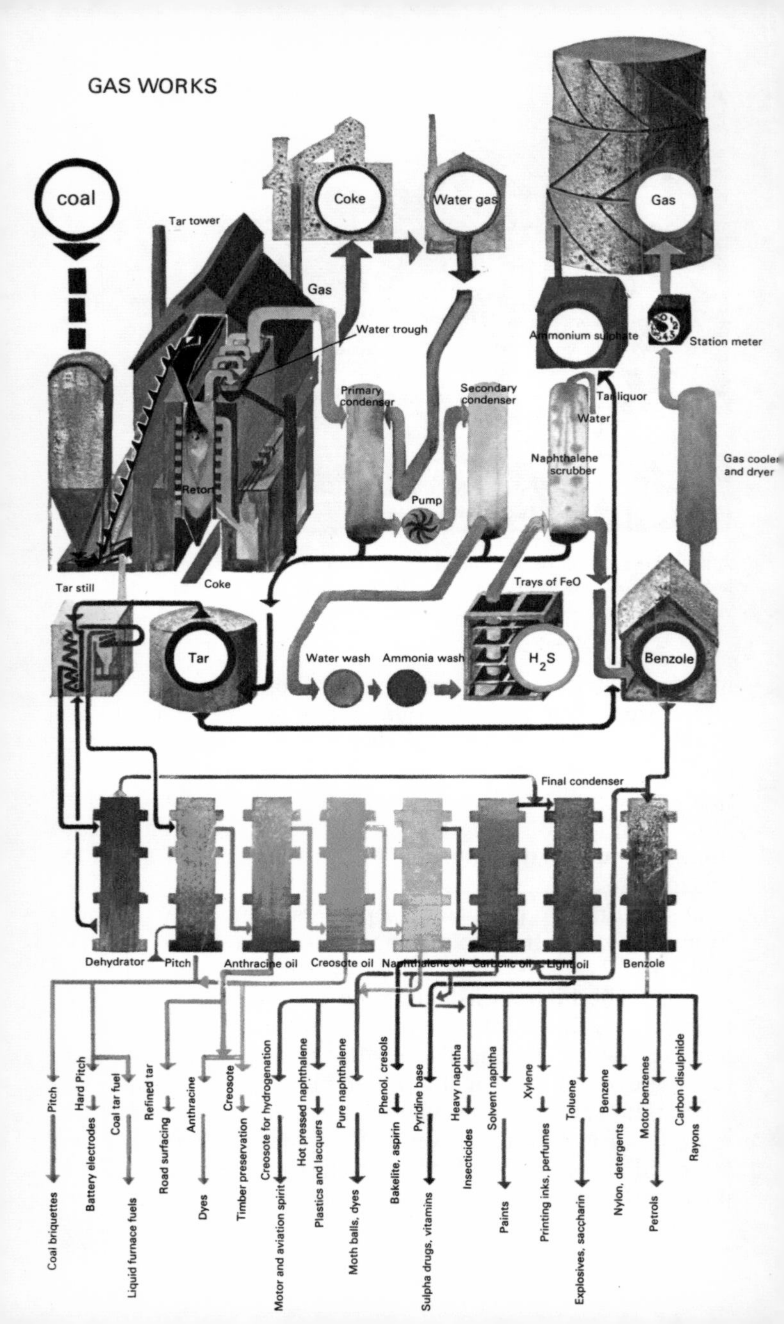

GAS WORKS
coal
Tar tower
Coke
Water gas
Gas
Gas
Water trough
Ammonium sulphate
Station meter
Primary condenser
Secondary condenser
Tar liquor
Water
Naphthalene scrubber
Gas cooler and dryer
Retort
Pump
Tar still
Coke
Trays of FeO
Tar
Water wash
Ammonia wash
H_2S
Benzole
Final condenser
Dehydrator
Pitch
Anthracine oil
Creosote oil
Naphthalene oil
Carbolic oil
Light oil
Benzole
Pitch
Coal briquettes
Hard Pitch
Battery electrodes
Coal tar fuel
Liquid furnace fuels
Refined tar
Road surfacing
Anthracine
Dyes
Creosote
Timber preservation
Creosote for hydrogenation
Motor and aviation spirit
Hot pressed naphthalene
Plastics and lacquers
Pure naphthalene
Moth balls, dyes
Phenol, cresols
Bakelite, aspirin
Pyridine base
Sulpha drugs, vitamins
Heavy naphtha
Insecticides
Solvent naphtha
Paints
Xylene
Printing inks, perfumes
Toluene
Explosives, saccharin
Benzene
Nylon, detergents
Motor benzenes
Petrols
Carbon disulphide
Rayons

OIL REFINERY

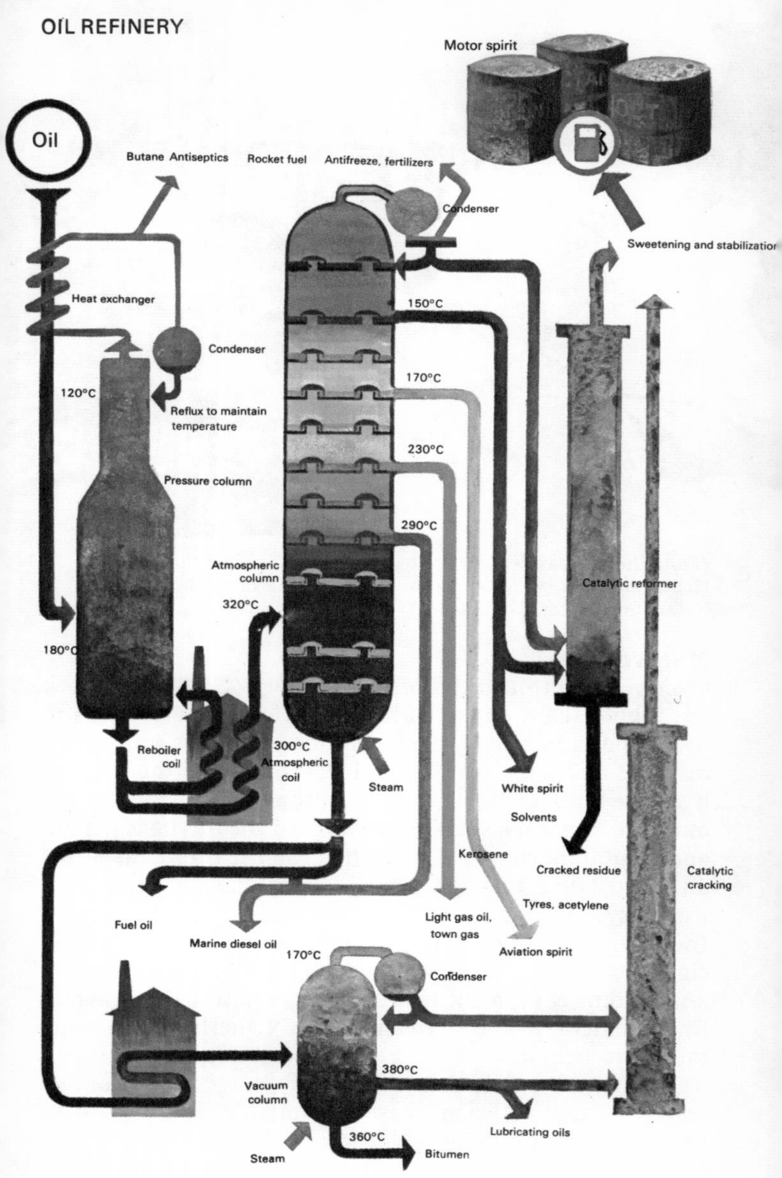

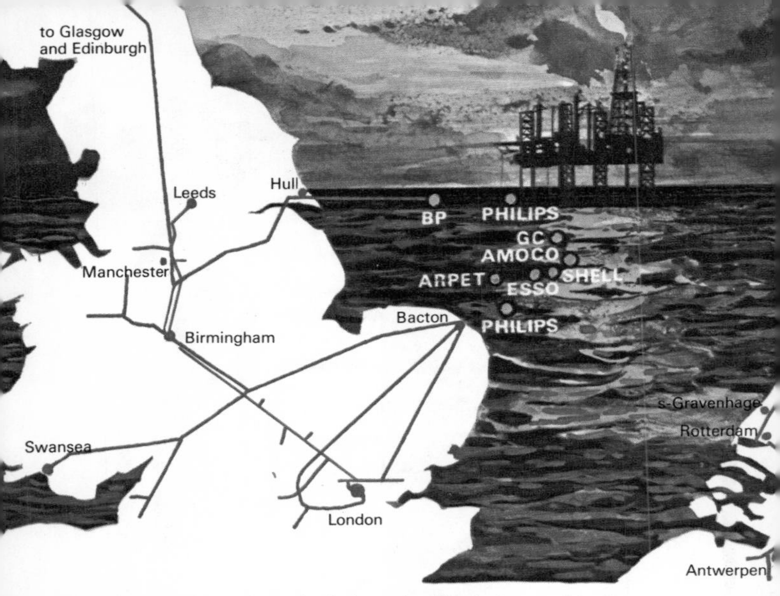

(Top) Natural gas deposits below the North Sea and in the Netherlands are being tapped to serve users of domestic gas in Britain and on the Continent.

Natural gas

Overlying deposits of oil or in separate pockets in the rock are found quantities of natural gas. Men drill for this in much the same way as they drill for oil, either on land or offshore under the sea. Natural gas consists of the paraffin hydrocarbons methane, ethane, propane and butane. (It may also contain a small percentage of the inert gas helium and is an important source of this gas.) It is used as a fuel and distributed through pipes like coal gas mains, or it is used as a source of organic chemicals and hydrogen. Individual gases may be liquefied and sold as fuel for cigarette lighters, caravan cookers, and portable cutting and welding equipment. The deposits of natural gas recently found under the North Sea and in the Netherlands are being tapped to help meet the increasing demands for fuel gases. The Slochteren field has the world's largest natural gas reserves, and sufficient gas is available for export to neighbouring countries.

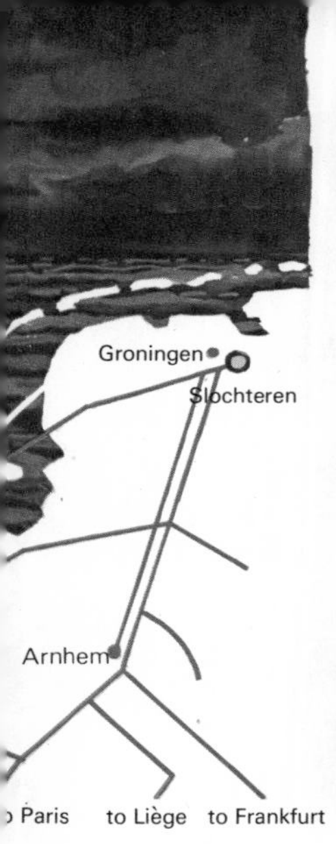

Blasting explosives are used in mining, quarrying and demolition work.

EXPLOSIVES

There are two main types of explosives: high explosives and propellant (or 'low') explosives. In warfare, high explosives are used in bombs, shells, mines, torpedoes and demolition charges. Their peaceful uses include demolition, blasting in mining, quarrying and roadmaking, and fireworks. Most of these uses also require explosive detonators. Propellant explosives are used in cartridges and solid-fuel rockets. The difference between the two types may be only one of degree and application. For example, gunpowder was used as a propellant in cannons and muskets but as a high explosive in grenades.

Most types of explosive can be considered as having two components (although they may both be embodied in a single chemical compound): a fuel–something which burns, and an oxidizer–a source of oxygen to make it burn very readily. Gunpowder is a mixture of carbon, sulphur and potassium nitrate. The carbon and sulphur burn and the potassium nitrate supplies the necessary oxygen. The products of combustion, carbon dioxide and sulphur dioxide, are gases. Their volume is large and the temperature of the reaction makes them expand even more. In the confined space of a gun breech this expansion forces the ball or bullet out of the barrel. In the confined space of an iron grenade case or a hole drilled in rock, the expansion blows the iron or rock to pieces. Single-substance explosives such as TNT (trinitrotoluene) contain oxidizable elements (carbon, hydrogen and nitrogen) and oxygen.

The story of explosives begins with gunpowder which was known to the Chinese in about the year

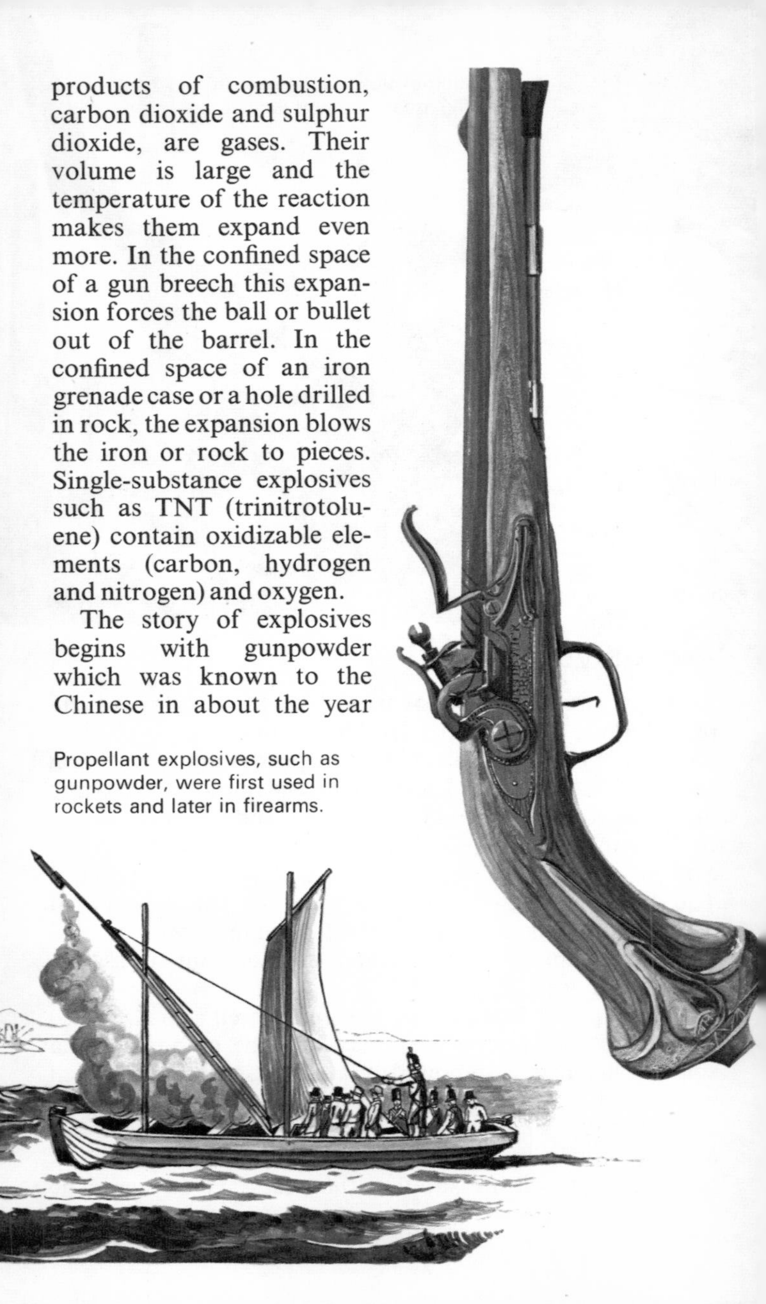

Propellant explosives, such as gunpowder, were first used in rockets and later in firearms.

1000 and used only as a propellant in rockets. Within a few hundred years its use had spread to Europe (it may have been reinvented there), where it was used almost exclusively in firearms and cannons. The Spanish conquest of South America was achieved with so few men because the Conquistadores had firearms and the Indians had not.

Gunpowder for guns was known as black powder (the individual grains were coated in graphite to prevent clumping). This was ignited by a train of fine powder lit by a spark generator such as the flintlock. The next development in explosives came with an attempt to do away with the priming powder and spark and to ignite the main charge directly. The solution, invented by William Forsythe, was the detonator. This was a pinch of very sensitive explosive, soon incorporated in the percussion cap, which exploded when struck by the gun's hammer and so set off the main gunpowder charge. The same system is used in firearms today except that the percussion cap is made part of the base of the cartridge. Detonators for blasting are fired electrically.

Chemically a detonator is an unstable heavy metal salt such as mercury fulminate, lead azide, or silver acetylide. They do not work by rapid oxidation (most contain no oxygen) but their molecules 'fall apart' when struck,

The development of the cartridge, with its self-contained propellant charge, gave rise to the machine gun.

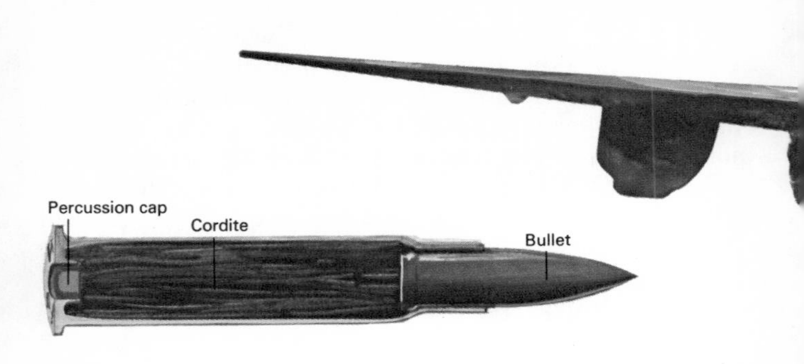

releasing the energy locked in their chemical bonds. They are easy to make but difficult and dangerous to handle.

The next great advance in explosives came with the invention of smokeless powders. The first of these, and still the commonest, was nitrocellulose. It was called guncotton because it was made by nitrating cotton waste. There soon followed many other such nitro compounds: trinitrotoluene (TNT), nitroglycerine, picric acid (trinitrophenol), and penta-erythritol tetranitrate (PETN). Nitroglycerine, or more strictly glyceryl trinitrate, is a liquid and could not be safely used until Alfred Nobel invented dynamite. This consists of an earthy substance resembling china clay which is impregnated with nitroglycerine.

The preparation of nitro explosives is basically simple. The parent organic compound, such as toluene, glycerine, or phenol, is nitrated with a mixture of concentrated nitric and sulphuric acids. More complex organic compounds can also be nitrated to make explosives. For instance the condensation product of ammonia and formaldehyde, called hexamethylenetetramine, can be nitrated to give cyclotrimethylene trinitramine. This explosive is known by one of the shortened names Cyclonite, Hexogen, or RDX.

For specific uses, high explosives are often 'blended'. For instance many bombs are filled with amatol, a mixture of TNT and ammonium nitrate. Torpedo warheads are filled with a mixture of TNT and RDX, called Torpex.

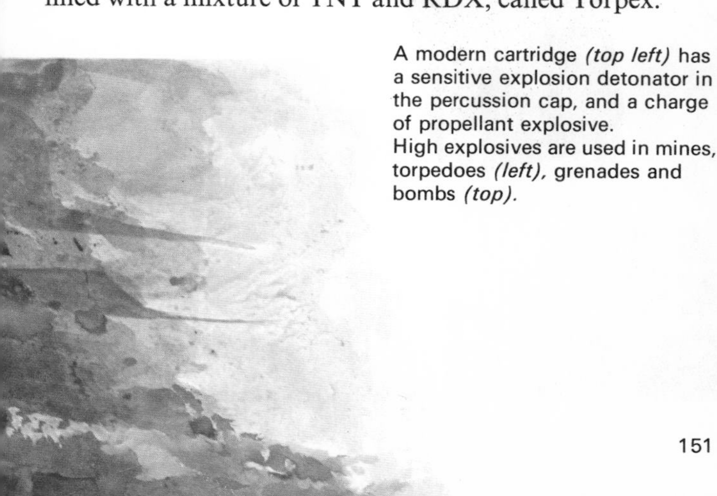

A modern cartridge *(top left)* has a sensitive explosion detonator in the percussion cap, and a charge of propellant explosive.
High explosives are used in mines, torpedoes *(left)*, grenades and bombs *(top)*.

DYES, DRUGS AND DETERGENTS

In this final section, the three remaining important chemical industries – all based on organic chemicals – have been grouped together.

Dyes

Originally dyes were made from natural products. Plant dyes such as woad, indigo and the chemist's litmus predominated, but shellfish (royal purple) and insects (cochineal) were also used. Then in 1856 William Perkin accidentally discovered the first aniline dye *mauvine,* which he made from coal tar when trying to synthesize the drug quinine. Other chemists turned to coal tar as a raw material and, in addition to producing more dyes, sorted out the organic chemistry of the compounds it contained. The synthetic dyes were cheaper and more light fast than the natural products. Also the best known natural dye, indigo, was synthesized from simple benzene compounds. Most synthetic dyes are benzene

The modern dyestuffs industry produces a wide range of colours for fabrics, plastics and inks.

or quinone compounds substituted with groups containing nitrogen.

Drugs

The development of the chemistry of drugs followed a similar pattern. At first there were only natural compounds, mostly derived from plants. These included alkaloids such as belladonna (atropine) from deadly nightshade, opium (morphine) from poppies, quinine from the cinchona tree, and cocaine, nicotine, caffeine and strychnine. They were called alkaloids because they give an alkaline reaction in solution. Gradually chemists worked out the complex structures of the drugs, and then they devised ways of synthesizing the substances from simple starting materials. Antibiotics, such as penicillin which occurs naturally in a mould, can also be made synthetically. Many of these syntheses involve dozens of separate reactions.

Penicillin and other antibiotics, many of which are derived from moulds, are made on a commercial scale in comparatively small batches and their action checked on cultures of bacteria.

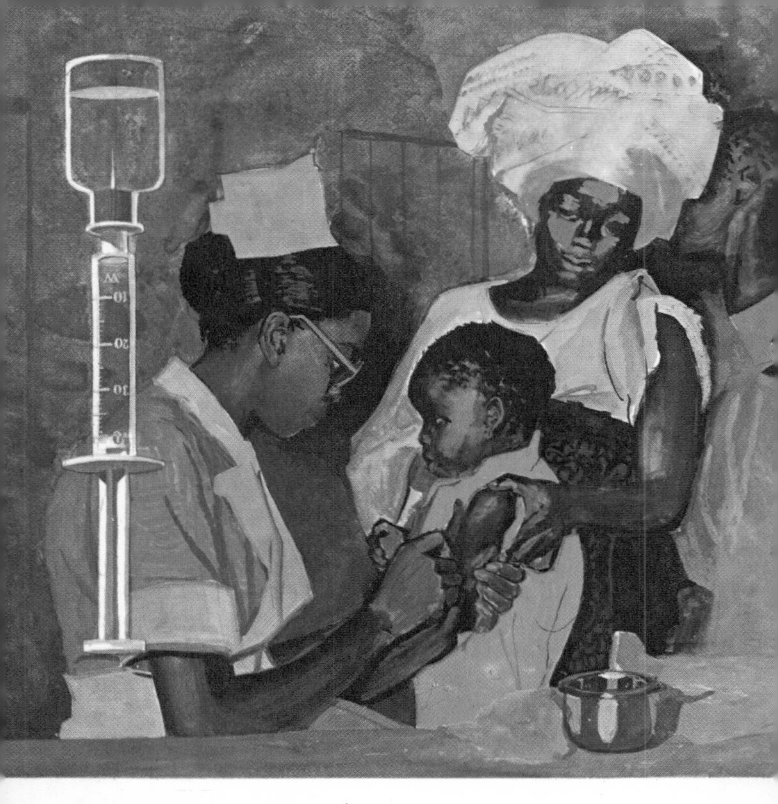

A sufficient supply of drugs such as antibiotics and sulphonamides has dramatically raised the standard of health.

Another class of drugs includes compounds that are entirely synthetic. One of the first of these was aspirin, which is the acetyl ester of a benzene compound called salicylic acid. Then in 1910 came salvarsan, an aromatic compound containing arsenic, which was used for treating syphilis. The sulphonamides were discovered in the late 1930s, the first British one being called M & B after its manufacturers May and Baker Ltd.

In recent years, biochemists have determined the structures of many of the body's chemical messengers, the hormones, and succeeded in synthesizing some of them.

Synthetic insulin is used to treat diabetes and sex hormones are used to make contraceptive pills. Many other artificial hormones are commonly used in medical treatment.

Detergents

For about 400 years, the textile industries and washer-women of the world had only soap to help them detach dirt from cloth. But soap is inefficient in hard water and in the acidic water found in textile factories. Research chemists developed a synthetic substitute–detergents. These compounds are sulphonates of paraffin hydrocarbons obtained from the oil refinery. The long aliphatic tail of the molecule is soluble in grease and the sulphonate head is soluble in water. Their calcium salts are also soluble, so hard water makes no scum with detergents. In detergent solution, grease smears on cloth fibres clump together as droplets which then emulsify with the solution.

In addition to their domestic uses, detergents have been employed to combat accidental pollution of the sea by crude oil.

BOOKS TO READ

The first two books on this list are at an elementary level; the remainder may be used for detailed reference in selected areas of the subject.

Explaining the Atom by S. Hecht and C. Rabinowitch. Gollancz, 1965.
Giant Molecules by Morris Kaufman. Aldus, 1968.
Theoretical Chemistry by S. Glasstone. Van Nostrand, 1944.
Valency and Molecular Structure by E. Cartmell and G. W. A. Fowles. Butterworth, 1966.
Chemical Elements and Their Compounds by N. V. Sidgwick. Oxford University Press, 1950.
Nuclear and Radiochemistry by G. Freidlander. Wiley, 271 28020 8.
Elements of Physical Chemistry by S. Glasstone and D. Lewis. Macmillan, 1963.
Reactions of Organic Compounds by W. J. Hickinbottom. Longmans, 582 44225 7.
Textbook of Micro and Semimicro Qualitative Inorganic Analysis by A. I. Vogel. Longmans, 582 44246 x.
Textbook of Quantitative Inorganic Analysis Including Elementary Instrumental Analysis by A. I. Vogel. Longmans, 582 44247 8.

INDEX

Page numbers in bold type refer to illustrations.

SOME OTHER TITLES IN THIS SERIES

Natural History

The Animal Kingdom
Animals of Australia & New Zealand
Animals of Southern Asia
A Guide to the Seashore
Bird Behaviour
Birds of Prey
Butterflies
Evolution of Life
Fishes of the World
Fossil Man
Life in the Sea

Gardening

Chrysanthemums
Garden Flowers
Garden Shrubs
House Plants

Popular Science

Astronomy
Atomic Energy
Computers at Work
The Earth
Electricity
Electronics

Arts

Antique Furniture
Architecture
Clocks and Watches
Glass for Collectors

General Information

Aircraft
Arms and Armour
Coins and Medals
Flags
Guns
Military Uniforms
Rockets and Missiles

Domestic Animals and Pets

Budgerigars
Cats
Dog Care
Dogs
Horses and Ponies

Domestic Science

Flower Arranging

History and Mythology

Archaeology
Discovery of
  Africa
  The American West
Australia
Japan
North America
South America